Philosophy of Science and
The Kyoto School

Bloomsbury Introductions to World Philosophies

Series Editor
Monika Kirloskar-Steinbach

Assistant Series Editor
Leah Kalmanson

Regional Editors
Nader El-Bizri, James Madaio, Sarah A. Mattice, Takeshi Morisato, Pascah Mungwini, Omar Rivera and Georgina Stewart

Bloomsbury Introductions to World Philosophies delivers primers reflecting exciting new developments in the trajectory of world philosophies. Instead of privileging a single philosophical approach as the basis of comparison, the series provides a platform for diverse philosophical perspectives to accommodate the different dimensions of cross-cultural philosophizing. While introducing thinkers, texts and themes emanating from different world philosophies, each book, in an imaginative and path-breaking way, makes clear how it departs from a conventional treatment of the subject matter.

Titles in the Series
A Practical Guide to World Philosophies, by Monika Kirloskar-Steinbach and Leah Kalmanson
Daya Krishna and Twentieth-Century Indian Philosophy, by Daniel Raveh
Māori Philosophy, by Georgina Tuari Stewart
Philosophy of Science and The Kyoto School, by Dean Anthony Brink

Philosophy of Science and The Kyoto School

An Introduction to Nishida Kitarō, Tanabe Hajime and Tosaka Jun

Dean Anthony Brink

BLOOMSBURY ACADEMIC
LONDON • NEW YORK • OXFORD • NEW DELHI • SYDNEY

BLOOMSBURY ACADEMIC
Bloomsbury Publishing Plc
50 Bedford Square, London, WC1B 3DP, UK
1385 Broadway, New York, NY 10018, USA
29 Earlsfort Terrace, Dublin 2, Ireland

BLOOMSBURY, BLOOMSBURY ACADEMIC and the Diana logo are trademarks of Bloomsbury Publishing Plc

First published in Great Britain 2021

Copyright © Dean Anthony Brink, 2021

Dean Anthony Brink has asserted his right under the Copyright, Designs and Patents Act, 1988, to be identified as Author of this work.

For legal purposes the Acknowledgements on p. xiii constitute an extension of this copyright page.

All rights reserved. No part of this publication may be reproduced or transmitted in any form or by any means, electronic or mechanical, including photocopying, recording, or any information storage or retrieval system, without prior permission in writing from the publishers.

Bloomsbury Publishing Plc does not have any control over, or responsibility for, any third-party websites referred to or in this book. All internet addresses given in this book were correct at the time of going to press. The author and publisher regret any inconvenience caused if addresses
have changed or sites have ceased to exist, but can
accept no responsibility for any such changes.

A catalogue record for this book is available from the British Library.

A catalog record for this book is available from the Library of Congress.

ISBN:	HB:	978-1-3501-4109-4
	PB:	978-1-3501-4110-0
	ePDF:	978-1-3501-4111-7
	eBook:	978-1-3501-4112-4

Series: Bloomsbury Introductions to World Philosophies

Typeset by RefineCatch Limited, Bungay, Suffolk

To find out more about our authors and books visit www.bloomsbury.com and sign up for our newsletters.

This book is dedicated to the fiction of our bodies in a quantum universe.

Contents

List of Illustrations	viii
List of Boxes	ix
Series Editor's Preface	x
Preface	xi
Acknowledgements	xiii
Note on the Translations	xv
List of Abbreviations	xvi

1	Introduction: Relativity and Quantum Physics in the Kyoto School	1
2	Nishida Philosophy, Place, Field and Quantum Phenomena	15
3	Mediation in Tanabe's Dialectical Vision of Competing Fields within Physics	67
4	Modern Physics, Space and Ideology in Tosaka Jun	111
5	What We Can Learn from the Kyoto School	135

Notes	151
Timeline of Kyoto School Activities in Relation to Physics	171
Glossary	177
Further Reading	181
Index	189

List of Illustrations

1.1	Scales in modern physics.	7
1.2	Physics overview.	10
2.1	Schematic situating the subject	60
2.2	A recurring schematic	62

List of Boxes

2.1	Systems	18
2.2	Percy Williams Bridgman	41
2.3	Standard measure	44
2.4	Topological psychology	58
2.5	Subject and object as series in Nishida	59
2.6	Legend to Nishida's last 'Schematic explanation'	61
3.1	Aether theory and the Michelson-Morley experiment	72
3.2	Einstein's special theory of relativity	73
3.3	Minkowski's world	77
3.4	Commutational centres	79
3.5	Newton's law of universal gravitation	79
3.6	Max Planck (1858–1947)	80
3.7	Max Planck's energy quanta and Planck constant	80
3.8	Arthur Holly Compton (1892–1962)	81
3.9	Newton's old emitter theory of light	82
3.10	Heisenberg's matrix mechanics	82
3.11	Bohr's theory of the structure of the atom	83
3.12	Erwin Schrödinger (1887–1961)	83
3.13	Louis de Broglie (1892–1987)	84
3.14	Tensor analysis and world tensors	94
3.15	Translational motion and rotation	95
3.16	Orthogonal	98
3.17	Dirac's quantum mechanics	99
3.18	Heisenberg's uncertainty principle	100
4.1	Space in Kant	115
4.2	Christiaan Huygens (1629–95), Georg Cantor (1845–1918), Luitzen Egbertus Jan Brouwer (1881–1966) and David Hilbert (1862–1943)	123
4.3	Nature dialectics or dialectics of nature	124

Series Editor's Preface

The introductions we include in the World Philosophies series take a single thinker, theme or text and provide a close reading of them. What defines the series is that these are likely to be people or traditions that you have not yet encountered in your study of philosophy. By choosing to include them you broaden your understanding of ideas about the self, knowledge and the world around us. Each book presents unexplored pathways into the study of world philosophies. Instead of privileging a single philosophical approach as the basis of comparison, each book accommodates the many different dimensions of cross-cultural philosophizing. While the choice of terms used by the individual volumes may indeed carry a local inflection, they encourage critical thinking about philosophical plurality. Each book strikes a balance between locality and globality.

Philosophy of Science and the Kyoto School: An Introduction to Nishida Kitarō, Tanabe Hajime and Tosaka Jun is a fine exercise in what Dean Brink calls a counter-hegemonic postcolonial gesture of differentiation. Brink foregrounds how Nishida Kitarō (1870–1945), Tanabe Hajime (1885–1962) and Tosaka Jun (1900–45) challenged the hegemonic force of European philosophy by situating it in their own local Japanese context, which itself is diverse and dynamic. By underscoring the use of contemporary discourses in modern physics in their philosophical work, Brink also paves the way for a counter-hegemonic reading of the Kyoto School itself. *Contra* the conventional interpretation, these prolific figures of the Kyoto School were not singularly focused on religious thought. Rather, they sought to develop a subtle, rich and fascinating account of scientific humanism grounded in local traditions.

Monika Kirloskar-Steinbach

Preface

World philosophies may be thought of as means of recognizing and realizing our innate hybridity. If one studies Hegel in France this truism might go unrealized, though when doing the same in Japan there has been a divide held between the local thought and the (Western) philosophy. The Kyoto School figures introduced here blurred this division by bringing to bear issues close to them in Japan – of whatever cultural or national origins, whether Buddhism, Marxism or physics – and articulated with great care their diverse orientations and passions for understanding the world physically and socially. With globalization such hybridity congeals in new norms for world philosophies, yet change in philosophy is like all academic pursuits glacial, as we say; we are just getting started. Now with the pandemic we are again somewhat sequestered in our divided worlds in times rife with international tension, and our task is to learn to love each other again (paraphrasing W. H. Auden). What does this have to do with the philosophy of science in Japan? To begin with, if national cultures retreat into their respective differences, science can provide points of controversy over interpretation that actually bring us closer to understanding each other's fundamental assumptions about the worlds in which we live. Our education suddenly includes the arts and sciences in an unusually technical sense that leaves little room for good intentions and opens many doors to observing how we each make use of opportunities afforded by scientific discoveries and implement them in our lives. Could scientific projects themselves lead embattled nations into cooperation that both could heal the colonial wounds of the imperialist capitalist past and offer democratic hope to people still struggling for basic human rights? Beginning perhaps with the Manhattan Project (to build the world's first atomic weapons) and similar projects in Germany and Japan, the possibilities for modern physics altering our world became gravely apparent. However, we have yet to join thorough studies in world philosophies in a sincere multicultural engagement with issues in the philosophy of science that are not superficially able to be folded into a (Western) master narrative. Scientific discoveries are far from having been smoothly integrated in the philosophy of physics today – though the discoveries themselves can be proven or disproven through a global network of peer review and repeatability of experiments. What can be explored through such work as is presented here is perhaps related to a complex of socio-political factors, including how ideologically sensitive issues and positions that science itself may support or challenge are modulated by the philosophy of science. The situation engaged by these Kyoto School writers offers an exceptional case of how philosophy of science may serve as a mediator of political, social, religious and scientific cultures in three diverse ways. In short, they each bring to bear on Western philosophy assumptions made in Japanese philosophy (especially by way of Buddhism) regarding ontological and epistemological starting points for matter itself as well as social constructs of various sorts (from individual perception to nationalist sentiment).

That we today are forced to retreat to our places of residence and the world seems at the cusp of a cold war, linked in part to the pandemic and in part to questions of technological and economic competition, makes the questions raised in the Kyoto School (on which a large fascinating and edifying discourse on war responsibility also exists) pertinent not only to scientific but also political cultures. As we negotiate cultural and political differences, how ideologies play roles within societies and in wars, and how the philosophy of science pushes our discourse within its ken towards issues of truth as matters of rethinking existing developments in science, one may find suggestions for solving pressing global problems by way of common yet distinct understandings of pollution and climate change, even the proliferation of destructive and coercive technologies by state and non-state actors. How can the philosophy of science contribute to the rethinking of a world of locales and hybridity such as is found in the example of the Kyoto School and promote a greater appreciation of the difficulty of attaining scientific commons in an age of misinformation? Always, it is the sheer – but never pure – intellectual beauty of situating modern physics as a driver of change in philosophy that may fascinate us and keep us reading.

Dean Brink, Hsinchu, Taiwan, 20 July 2020

Acknowledgements

This book owes its life to the encouragement of the series editor who today is promoting world philosophies more than any other philosopher, Monika Kirloskar-Steinbach at the University of Konstanz, who carefully read and provided invaluable comments that shaped this book into its present form. The preliminary research that led to such a book began at a roundtable exploring science and culture hosted by Iris van der Tuin at Utrecht University in 2017, though my work on Nishida Kitarō and Tosaka Jun began long ago at the University of Chicago, where Tetsuo Najita first pointed me to the work of Emmanuel Levinas, a lead I did not investigate until my wife, Joy Shihyi Huang, a theatre scholar sometimes featuring his work, created the possibility for exploring it further. Karen Barad, who I have never met, stimulated me to read widely in physics in order to understand what decisions led her to exhibit such conviction as to reduce questions in physics to the work of Niels Bohr and then elide it with the deconstructive framings of Jacques Derrida; recent work in quantum physics (which by its very nature is theoretical) bodes ill for this direction, as is explored in the last chapter. Of course, it is to the three philosophers who are introduced here that I owe the greatest debt of gratitude, for it is their unstinting dedication to philosophy in an age when originality and the risk-taking in offering new frames of coherency led to great intellectual feats and challenges. I also owe a debt of gratitude to the anonymous reader who went beyond the call of duty with immense generosity in making exquisite suggestions for the final tweaks and refinements that hopefully render this book of utmost use to readers.

More than anyone in the field of physics, Rossella Lupacchini at the University of Bologna patiently guided me through controversies in quantum physics that proved to be extremely clarifying. Makoto Katsumori, a Bohr specialist, and Jacynthe Tremblay, who works on physics in Nishida (in French), each provided camaraderie and support, and lent me confidence to pursue this forgotten corner of Japanese philosophy and restore it to its proper influential role in the Kyoto School. In the course of writing this book, I have also benefited from conversations with numerous scholars, young and old, including especially Morisato Takeshi, John Maraldo, Jim Heisig, Edward George McDougall, Kyle Shuttleworth, Hans Peter Liederbach, Roman Paşca, Graham Mayeda and Lucas Nascimento Machado. I thank Becky Holland and Zoë Jellicoe at Bloomsbury for seeing my work through the many steps toward realization with great care.

Minute portions of the Introduction and Chapter 5.1 first appeared in a preliminary study, 'Quantum Dialectics', in *Philosophy Today* 63, no. 4, an issue on a special theme, 'New Concepts for Materialism', edited by Iris van der Tuin and Adam Nocek (Fall 2019). The resources for attaining the complete works and other secondary sources needed to complete this book as well as attend Japanese philosophy conferences in Germany and Japan were generously provided by research grants from the Ministry of Science and Technology, Taiwan. I also thank the National Chiao Tung University Library for maintaining a rich collection on modern physics and for their speedy interlibrary loans.

Note on the Translations

Translation style in Japanese philosophy is highly contested. Some critics may prefer a literal word-for-word rendering of both syntax and semantics intact as if to render a strictly parallel linguistic universe, while others find such limitations to result in stilted English and thus opt for taking stylistic liberties in the interest of readability. I aim for both exactitude and readability, shifting parts of speech when the meaning of the primary relationships indicated have already been made clear in a preceding clause, freely omitting repetitious words and adding connectives, and converting the copula into more active (though static) verbs that are preferable in English. Such an approach is more precise, since it will be English, expected to be English, and allow for subtler distinctions and connotations. Although in English it is customary to avoid repeating words (as it can quickly become grating), I have tried to be consistent in the case of translating key concepts, many of which are listed in the glossary and sometimes annotated in the notes or discussed in boxes. For some long sentences, vital information is moved to an adjacent clause (or even sentence, though rarely) with the primary aim being to approach idiomatic English faithful to the relations presented in the original. Of course long sentences are divided and short ones combined (one of Jay Rubin's keys to translation) and paragraph breaks (eschewed by Tanabe) are inserted as needed, all in the interest of readability in English. In all renderings, precision in a parallel universe is the prime directive.

In terms of translating antiquated concepts in physics, I have updated terms which have been superseded by newer coinages of concepts whose meaning has not drastically changed in an attempt to reflect parallel usages in English; however, I have remained faithful to the original contexts and translations of now-obsolete terms of historical interest, making annotations as needed.

Some of the translations are complete chapters from Nishida and Tanabe, or extensive selections from one of their works. Instead of breaking up the text, commentary is inserted where deemed helpful in paragraphs rendered in a distinct font. This strategy is designed to allow readers to access the continuity and intentions of the authors while still benefiting from guidance through sometimes dense and difficult passages and highly original ideas.

In Japanese, names are written last name then first name in the text; to readers familiar with Japanese, to present them in another order may induce migraines. Thus, I follow Japanese usage in the first mention and then use only the last name.

Any shortcomings are of course my own responsibility. With the aim of improving future editions, I always welcome comments and suggestions.

List of Abbreviations

THz   Tanabe, Hajime (田辺 元), *Tanabe Hajime zenshū* (Complete Works of Tanabe Hajime), 15 volumes, Tokyo: Chikuma Shobō, 1963–4.
TJz   Tosaka, Jun (戸坂 潤), *Tosaka Jun zenshū* (Complete Works of Tosaka Jun), 5 volumes, Tokyo: Keisōshobō (勁草書房), 1966–7.
NKz   Nishida, Kitarō (西田 幾多郎), *Nishida Kitarō zenshū* (Complete Works of Nishida Kitarō), 19 volumes, Tokyo: Iwanami, 1978–9.

1

Introduction: Relativity and Quantum Physics in the Kyoto School

Philosophy of Science and The Kyoto School explores the implications of new developments in physics and related dialogues among Japan's leading philosophers in the first half of the twentieth century. The scope is mostly confined to issues in physics as understood at the time of writing the works presented here in English translation. The philosophers each have their own interests that lead them to read across a range of philosophical disciplines, citing the Greeks as well as the latest journals and books in a variety of languages, primarily German and English. Each philosopher approaches questions central to the philosophy of modern physics by entertaining shifts in the framing premises usually assumed in Western philosophy. Some of their assumptions are shared, but each of them develops their understanding in separate directions. These philosophers engage in one of the most exciting and rather daunting innovations of a world philosophy, now known as the Kyoto School.

For clarity's sake, one can say that for someone born in the early Meiji Period (1868–1912), philosophy (*tetsugaku* 哲学), as established in the West, was treated as more advanced and superseding existing published treatises on controversies of the times in Japan. However, as all disciplines become technological forms of engagement, one might expect philosophy, like chemistry or shipbuilding, to be able to be mastered through objective means of presentation and communication of methods and results, just as approaches to chemistry or shipbuilding in Japan would apply in France and vice versa. Yet philosophy is somewhat different in that it is ultimately informed by questions coined in local languages that are translated between and incorporated into international discourses. This is not only true of Japanese incorporating Chinese thought and later German thought, but of Chinese incorporating Indian thought and Germans incorporating Greek thought. Though the localities in Greece and India (and China) may be abstracted and the use in a new location seem to transcend them, ultimately to philosophize requires localization. The subject writing philosophically, as it were, requires some recognition of efficacy, of authenticity of voice or manifest motivation that somehow organically emerges from a locale and clarifies whatever issue is raised. The worth of the writing may rest on some form of relevancy to some situation, real or imagined, but rooted in a writer's milieu.

To write as someone coming of age in Meiji Japan but then assuming the tradition of Western philosophy to an extent would be to present oneself as indistinguishable from someone writing in France or Germany, would amount to an academic exercise in mimicry or rote learning. Indeed, early work in Japan that merely introduced Western philosophy was the norm, before Nishida. While Japan had its own philosophy, usually referred to as thought (*shisō* 思想), based variously on Buddhist, Confucian and National Studies (*kokugaku* 国学) discourses, it even today has yet to be brought into conversation with Western philosophy to the degree that a world philosophical scope could be assumed to constitute a commons. Japanese *thought* has existed for over a millennium, partially in dialogue with Chinese and Indian religious and secular thought and philosophy; however, Japanese *philosophy* is a problematic term because for over a century 'philosophy' has been used to refer to the more Eurocentric Western philosophy in contradistinction from Japanese thought, and the very authenticity or relevance of this philosophy as Western philosophy in Japan is challenged by a long-standing more home-grown dialogue with Indian, Chinese and Western philosophical traditions, which exists primarily as historical studies that sometimes stir contemporary controversies (especially when touching upon issues of interest to nationalists). Thus, due to historical contingencies, various Neo-Confucian and National Studies scholarship dating from pre-modern times to the present are not usually categorized as 'philosophy' but rather as 'thought'.

Interestingly, Kyoto School philosophers and their contemporaries are found in both the philosophy and thought sections of bookstores today. Within Japan, to be a 'philosopher' (*tetusgaku-sha* 哲学者) assumes an engagement with Western philosophy, which indeed characterizes Kyoto School work in the early half of the twentieth century. To be engaged in thought may include the more abstract writings in literary theory, cultural studies and sociology, and to engage in contemporary debates that includes the freer writings by philosophers in the academy is to play the role of a 'thinker' (*shisō-ka* 思想家). To put it simply, what counts as most influential in Japan – philosophy (*tetsugaku* 哲学) or thought (*shisō* 思想) – may change according to both nationalist and international cultural capital and realpolitik power configurations, so neither can be said to be above the other. Just as Japan disavowed affiliation with Chinese thought and culture in its turn to Western technology in the 1850s, it could again ally with China as the latter may attain the status of a new hegemon in the region. Time will tell. This speculation also suggests how future directions in world philosophies too will likely depend on such unforeseeable contingencies, and could entail some form of decentralization of European thought. For instance, there are many forgotten writings in Japanese and Chinese Neo-Confucian philosophy or thought that date back centuries and yet form alternative perspectives on language, governance, ethics and human relations that could produce global interest in philosophy.

In any case, how subjects relate to objects, intuit fields of relations and questions of space are formulated by two central figures in the Kyoto School of philosophy, Nishida Kitarō (1870–1945) and Tanabe Hajime (1885–1962) as well as one Kyoto School student, turned critic, Tosaka Jun (1900–45) in ways that challenge Western philosophy on its own terms yet with local assumptions. Rather than treat Japanese philosophy as Japanese, per se, one can perhaps more clearly see it as presenting a Japanese Western philosophy just as one has a Dutch or Finnish Western philosophy. One may call it world philosophy or assume a world rather than Western hegemony in simply using *philosophy* to describe it. After all, that is what it is called in Japanese (哲学). An underlying distinction that might be said to have founded (or grounded) the Kyoto School may be attributed to how Nishida philosophy (*Nishida tetsugaku*) substituted an assumption of an unchanging being – prevalent since Parmenides (early fifth century BCE) – with one of nothingness or nothing (*mu* 無). This nothingness may be understood as less a nothingness within a field of *a priori* being than as emphasis on what primordially emerges out of a given void (or emptiness, *kū* 空, the Buddhist translation from Sanskrit of *Śūnyatā*) rather than assuming being as a starting point. Nothing is not an absolute absence of being but rather may be thought of as a blurring of being in an originary undifferentiatedness out of which questions of knowledge and being may be freshly considered. *Nothing*, in English, has the unfortunate tendency to invoke a binary opposition with being. Nishida attempted to construct a philosophical system based on alternative ontological and epistemological orientations of human subjects. To retain the framework of being as a privileged starting point would be intellectually dishonest in treating Nishida (and indeed how it ripples through the thought of the Kyoto School), once one has noticed this distinction. Thus, void, with its sense of a non-existing singularity or undifferentiatedness, is helpful in situating 'nothing(ness)' in Nishida and others influenced by him. What makes the philosophy of science of the Kyoto School so fascinating is precisely how this alternative Asian ontological assumption has found its affirmation in modern physics, providing opportunities for rethinking a range of issues touching on political philosophy as well as the philosophy of the foundations of physics.

Postcolonial aspects and implication for world philosophies

This focus on local space in a dialectics of void and being as well as relational empirical interaction helps one understand the significance of a dialectics of empty space in our understanding of space and objects in quantum theory. However tenuous a relation to human presence, Tanabe's engagement with the forefront of developments in new quantum physics leads him to dissolve the very premises for postulating spaces; he implicitly links it with Nishida philosophy (in these earlier works), which in part inspired his own thought. Here we may begin to situate their work in light of Emmanual Levinas's writings on *empty space* and as forming overlooked contributions to the philosophy of science. This framing helps us understand the postcolonial dimension of modern physics in terms of its contested implications and opportunities it offers (by no means as a coherent whole) for rethinking orders of knowledge and how we coexist in physical situations. By reversing the usual preference for light over darkness, Levinas recasts the fullness of a space illuminated by the rational objectification of an Enlightened society as *removing* the mediator *in relation to others* from the world and, instead, proffering merely an autonomous system that provides a negation of interbeing interaction and being. Light and the categories of Enlightenment writ absolute form alienating homogenous spaces and 'empty space [becomes] the condition for' relationality, with local spaces ensuring 'the condition for the *lateral* signification of things within the same'.[1] Moreover, not only Nishida but Levinas too would recognize the importance of *intuition* to mediate relations between things, people and non-humans. The full hubris of embracing the Enlightenment removes the immediacy of relations by 'driving out the shadows; it empties space',[2] or, as Nishida or Tanabe might say, the Enlightenment denies a radical dialectics of space. One now finds only beables (see Chapter 5), that is, anything that can possibly be, within a spaceless spatiality (abstract configuration spaces).

Nishida, along with Tanabe, can be understood today as extending our understanding of space in precisely Levinas's sense. One cannot ignore the postcolonial implications of this attempt and strategy. What can be discerned is that Nishida in effect engages in a counter-hegemonic postcolonial gesture of differentiation by introducing *nothingness* (*mu*) into what was and remains being-oriented 'Western' philosophy.[3] Nishida should be recognized as the non-Western philosopher par excellence who transformed Western philosophy into world philosophy to the benefit of all. He deployed emptiness as nothingness in order to diffuse space as a site of mastery and in ways that may overcome pseudo-colonial relation with an other.

What one can trace in terms of a dialectics in light of modern physics is a movement toward recognition of the *local* in both the foundational physics of our time, quantum physics, which Tanabe explores *as a dialectic*, and relativistic emplacement, in Nishida's sense, of human-world dialectics situating humanity as interacting mediators *among* things as well *as* things in a field of things (inspired in part significantly by relativity theory and, to a lesser extent, by quantum theory). In conclusion, the possibility of a dialectic of *empty space*, given that according to quantum theory space cannot be empty (at any time wave-particles have a probability of appearing anywhere), may for convenience be framed as forming a dialectic with relativistic space-time and all its classical trappings. Such an appreciation of possible states of space and matter would seem helpful in framing radically new materialist concerns, but is only a preliminary step underscoring our particular contexts of inquiry.

Book overview

Through summary and translation of essays and passages from works that have yet to appear in English, each chapter explores the specific interests of Nishida, Tanabe and Tosaka in relation to questions in the philosophy of science. Chapter 2 includes an earlier work by Nishida that situates his thought in relation to leading philosophers read in his time in order to elaborate what he understands by active intuition, *poiesis* (making) and will in relation to a physical place – key concepts from this period forward. It ends with a work engaging problems in modern physics, focused on the inspiration he drew from Bridgman's Operationalism, and includes related explanations of the symbolic language and schematics that Nishida invented to explain his system as it matured in his late work. Chapter 3 includes two representative samples of Tanabe's writings on modern physics, particularly the challenges and opportunities presented by quantum mechanics. The two translations represent his innovative reconfiguration of problems of dialectics and history to variously competing epistemes, notably quantum physics and classical physics, and even maps within his inflection of a Kyoto School metaphysics to what Dirac called void-holes (which we today associate with the discovery of antimatter and the proton). Chapter 4 explores Tosaka's own interpretation of how Kantian approaches to space and time influence his thought even as a Marxist, and how he situated the sort of epistemic dialectics that Tanabe also explored, each with entirely different conclusions. The final chapter connects these Kyoto School writers to contemporary issues in the philosophy of science (notably quantum physics) and more speculative questions relating to post-humanism, especially issues concerning how we situate human and non-human forms of agency, interaction and changing ecologies. Focusing on physics in particular, the legacies of Kant and Hegel and the impact of Neo-Kantian and Marxist materialism on the trajectories of these three Japanese philosophers are explored in the last two chapters. That Tanabe and Tosaka constructed their thought in conjunction with and in breaking from the founder, Nishida, can be explained in part not only by noting the abstract Hegelian and Kantian predilections in their thought, but due to their distinct relationships with the philosophy of science – a topic rarely breached, despite the large quantities of writing on science in each. Nishida, an extremely original philosopher, in part found reinforcement of his positions reflected in the work of Niels Bohr (1885–1962), a pioneer in quantum physics, and in part expropriated a narrow band of writing by a rather eccentric physicist, P. W. Bridgman (1882–1961), a specialist in high pressure physics, who also wrote popular works on relativity and quantum physics. Tanabe, for his part, read more widely in physics and the philosophy of physics to explore subtler issues in the philosophy of science in light of his own philosophical agenda, notably turning to Paul Dirac (1902–1984), known for breakthrough synthesis of relativity theory and quantum physics and invention of a broad range of concepts that are central to the history of quantum mechanics.

This book focuses on writings engaging issues in the philosophy of science during the revolution in physics precipitated by Einstein's special relativity theory (1905) and quantum physics, a new branch of physics studying the quantification of matter inclusive of light and electricity. These developments in physics were as important to contemporary philosophy as Isaac Newton's (1643–1727) laws of classical mechanics were to Immanuel Kant (1724–1804), whose system of framing concepts (such as space and time) and transcendental categories derive from selective appropriations from Newton.[4] Readers will come away with a solid grasp of the foundational philosophies developed in this period of extraordinary effort and creativity seldom found today in Japan or elsewhere. The founder of what we today call the Kyoto School of philosophy, Nishida, distinguishes himself from immediately preceding thinkers who were interpreting and translating Western philosophy using various Confucian and Buddhist concepts with which they were familiar.[5] Nishida began his university studies in 1891, just as the foundation of scholarship and international exchange had reached a point that an intelligent and ambitious student could have access to texts both in translation and in German, English and French. When Tanabe studied in Germany for two years (1922–4), Heidegger served as his tutor at the very time he was lecturing on materials that would appear in *Being and Time* (*Sein und Zeit*, 1927); one can see not simply an influence traceable to specific references, but a mode of thought and framing of existence and life in Heideggerian terms in general, but also as being redefined and rejected. Tosaka was a brilliant student and critic empowered by a charismatic swagger and the confidence he placed in the then popular communist movement; he maintained his convictions though it led to his early death in prison.

The three figures are presented in chronological order, chosen for inclusion due their general prominence and degrees of engagement with modern physics. Although there are other works by them that treat biological sciences, also within the purview of the philosophy of science, physics provides a common thread here. Each of them writes on long-standing issues in the philosophy of physics, including conceptions of matter, form, metaphysics, phenomenology, cognition, representation, ontology, epistemology, time and space and *a priori* transcendental frames and categories.

Modern physics

One point that needs to be made early on is that modern physics in Japan is continuous with modern physics anywhere.[6] What is specific to Japan are the cultures within which science is shaped and – in the case of key Kyoto School philosophers Nishida and Tanabe, including one of their most outspoken students and later critics, Tosaka Jun – used to support their own systematic explorations of metaphysical issues in phenomenology, ontology and epistemology. They also variously affirm and contest various socio-political

Introduction: Relativity and Quantum Physics in the Kyoto School   7

positioning of subjects in the world and within Japan in an age still defined by national power and internationally competing and mutually constituting colonial hierarchies, which as a whole become a hegemonic world order based on their balance of power and influence. It was in part by leveraging Nishida's status as the greatest philosopher of modern Japan that the likes of Einstein and Bohr were invited to Japan to give talks. The Japanese physicist Hideki Yukawa (1907–81) would himself play an important role in debates on quantum physics, with his prediction in 1935 of elementary particles called mesons, for which he received a Nobel Prize in 1949, further enhancing the visibility of Japan in the world of modern physics.

Like studies of phenomenology, ontology and epistemology in philosophy, modern physics problematizes how we codify in symbolic formulae and natural language the detection of matter and postulated properties and the scale of our awareness. Are our laboratories for observation confined to particle accelerators (atom smashers) or do they include various telescopes reaching the edge of visible time? Both of them approach the origins of matter and the nature of matter, albeit on different scales. In many ways, like anything modified by the word modern, modern physics is characterized by new frames for understanding matter and how our perception of it is bound up in our placement in relation to it. Two directions dominate: relativity theories and theories explaining quantum phenomena (see Figure 1.1).

To approach the Kyoto School without giving prominence to their writings on the hard sciences is to ignore the role of personal preferences on the part of Nishida, Tanabe and Tosaka for scientific thought of various sorts (including Soviet-approved approaches to science, for instance, in Tosaka). One may even say it is to project a Romantic (or even Orientalist) assumption

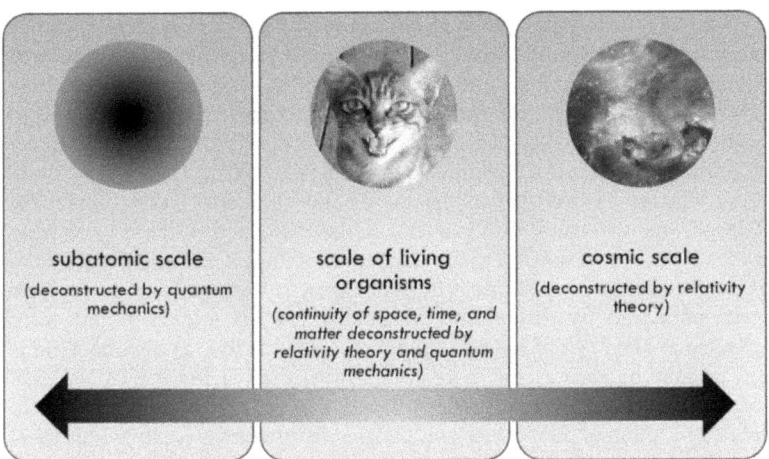

Figure 1.1 Scales in modern physics.

that Japanese philosophers are as much enamoured by discourse on religion as the Western scholars who dwell on it in Japan. In fact, mathematics and physics become explicit and far more prevalent topics for them themselves. Yet, their prominence has been largely ignored for decades as the Kyoto School emerged in scholarship in Western languages as predicated more on Christian dialogues with Buddhism than in conversation with discourse on Minkowski and Einstein, or Dirac and Bohr.[7] These three prolific Japanese philosophers engaged issues in science in an age when new discoveries in physics were being made in an ongoing dialogue among physicists throughout the first few decades of the twentieth century, precisely the time when they were most active. Thus, this collection of translated chapters and shorter excerpts from writings by Nishida, Tanabe and Tosaka suggests new opportunities for further study.

The primary questions raised by them concern relativity and quantum physics. This book assumes no previous background in science nor does it aim to present a comprehensive introduction to questions raised in the intense debates among physicists in the wake of the decline of classical physics as new concepts and experiments would demythologize it. Ongoing scientific findings and arguments published within the field of physics and the philosophy of science reflect a common international discourse vetted through peer-review and of course including Japanese scientists. This global scope forms a backdrop for philosophical debates in Japan that are shaped by issues in philosophy specific to Japan and discourse in Japanese. This book introduces issues concerning how the best minds in Japanese philosophy who happened to be drawn to questions in science each have distinct orientations that inform their engagements. Throughout their oeuvres, each approaches and integrates such questions into their thought, with each drawing fascinating and distinctive conclusions.

Debates on the significance of physics within the philosophy of science necessary entail a complicated situating that can be clarified as questions of ontology (of being, nature and object), questions of epistemology (of how knowledge is formed and ordered) and sometimes questions related to political philosophy (to idealism, dialectical materialism), as well as many more issues. One recurring figure in this regard is Martin Heidegger (1889–1976), who was visited by many Japanese philosophers, including Tanabe (in 1922, originally to study with Husserl), in the 1920s, and who most famously (and at times notoriously) characterized meaning in life in terms of the national context in which one is born or 'thrown'.[8] Though Tanabe may have been influenced by this ethnic essentialism in his middle-period work collected as *The Logic of Species* (*Shu no ronri* 種の論理),[9] as a product of his age, he may be seen as far more complicated than Heidegger (a National Socialist). This complication reflects a period in world history that was still defined by colonial hierarchies and racializing discourses (such as eugenics) as well as Darwinian and Spencerian depictions of nations as being in competition of a sort not only reflected in military might but in putatively fixed national racial orders.¹⁰ This period was particularly vulnerable to fallacies eliding national greatness with technological development. Though some Europeans were the first to modernize and colonize, Japan avoided colonization by modernizing itself very quickly, from 1860 to its victories in the First Sino-Japanese War (1895–6) and the Russo-Japanese War (1904–5). It is in the late stage of this context of modernizing national rivalries reshaping the world that not only did German National Socialism arise, backed by Heidegger, but Japanism, which has led to debates over the extent Tanabe in particular supported Japanese militarism and expansionism. In any case, issues surrounding how to define a *Japanese* philosophy, when philosophy's very definition relies on a European cultural frame, has led to complications. Though postcolonial studies has largely painted a binary between Europe and the rest, the Japanese Empire displaced a Eurocentric hegemony, leaving many newly problematized issues that include how racialization – reproduced codifications and associations invoking hierarchical valuations of ethnic groups – figured in Tanabe's *The Logic of Species* as an attempt to overcome this sine qua non of colonial power formation.¹¹ After all, it was the imposition of unequal treaties and extraterritorial rights for Europeans that fuelled Japanese nationalism and the determination to be treated equally in the world; however, even while Japanese modernization itself inspired global colonial independence movements, Japanese colonial ambitions themselves led to a racializing regime that naturalized a hierarchy within the empire that placed Japanese 'naturally' at the top.

Nishida was by no means a nationalist by most measures, but he understood in his own way that as an individual in Japan his experiences indeed did not prepare him to accede to the cultural assumptions found in Europe and reflected in European philosophy. As someone versed in Chinese and Japanese classics and especially as having dedicated years to Buddhist meditation, he had been prepared in life for a creative engagement with philosophy from a distinct, necessarily local and cosmopolitan perspective. This cosmopolitanism rather than nationalism allowed him to develop his own philosophy, now known as Nishida Philosophy, that formed the cornerstone of assumptions by which the Kyoto School of philosophy would flourish, expanding to include Tanabe Philosophy and many other prominent individuals even to this day.

Kyoto School engagements with relativity and quantum physics in Japan (1910–50)

Modern science has close ties to philosophy since the implications of one influences the other, calling for constant mutual readjustment of epistemological frames for ontological givens, and shifting preferences for fixed forms of *a priori* axiomatizations or empirically framed tentative formulations in the long view of scientific understanding. How to situate the implications of experimental findings in a world of modern physics that abandons space-time and even the solidity of everyday matter, not to mention matter itself? It was during the very period in which these three Japanese philosophers wrote that concepts and relationships between forms of matter (particles or waves) and relationality were undermining classical determinations of observed objects and observing subjects, as well as concepts such as causality and even how to define scopes of inquiry. Most of this book will treat issues in relativity theory and quantum mechanics, which emerged as the driving force of modern physics in its displacement and discrediting of classical physics (see Figure 1.2).

Einstein's relativity theory is often mentioned in the writings of these three, though put to different uses. How Nishida, Tanabe and Tosaka each quite distinctly incorporated discussions of physics in their various engagements with systems of thought and contexts (in and outside laboratories) may be expected to explore questions concerning social relevancy far beyond physics. For instance, can we entirely recast all we know of their works by overlooking the role of science in their works, often at the expense of overemphasizing the comparatively fewer writings on religion in Nishida and Tanabe?[12] But, on the other hand, might it be justified to charge the focus on philosophy of science here – by Nishida, Tanabe and Tosaka as well as in this book itself as a reading of the Kyoto School – with 'scientism', the privileging of science as something 'exacting' and 'objective' that thus provides new fertile and indubitable foundations for establishing their thought? Of course the very question is complicated by physics in the twentieth century, which in effect holds a half-transparent mirror up to the

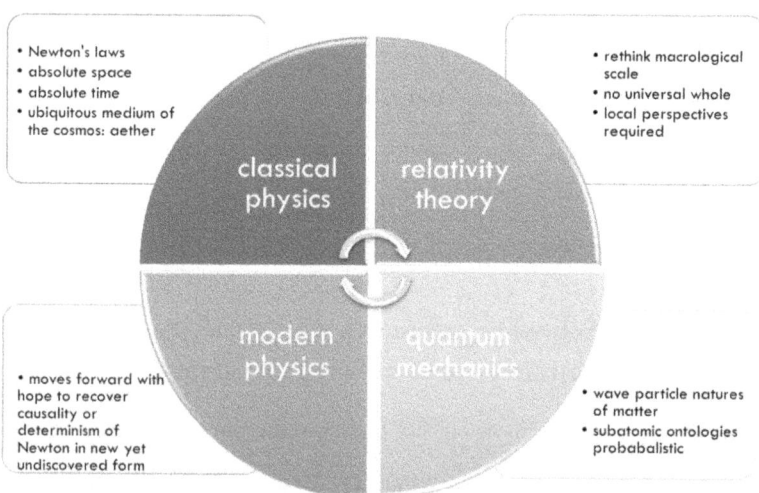

Figure 1.2 Physics overview.

human gazing on nature, dissolving both human ascendency and objective knowability of the world. Yet, we still may ask what interdisciplinary and interdiscursive approaches present possible avenues for a philosopher in the contexts of Nishida, Tanabe and Tosaka, each of a slightly later generation with different dispositions and expectations. In the end, we are obliged, I think, to assess the depth of both the engagement with philosophy as philosophy (in relation to other philosophers and thought in general) and the role science plays in their philosophical approaches to it. That is, we may productively gauge the depth of their understanding of issues in science that they discuss in light of the context in which they lived and, if possible, in light of the possibility of establishing their relevancy today, with comments relating to later discoveries and discourses in the philosophy of science (a topic raised in the final chapter).

Let us begin with a general assumption regarding Nishida's founding of the Kyoto School, the first home-grown philosophical system engaging Western philosophy on Japanese terms. He begins with a demonstrated interest in shifting Western dialectics of subject and object (or self and nature) in an assumed world of being, instead, to a dialectics premised on nothing (*mu*). Much has been written on how it is born out of his studies of Zen Buddhist thought in which a void prevails as an undifferentiated substrate defining the emergence of the world and self (and selflessness), but less on the postcolonial tensions that are equally relevant. Though Japan was not colonized, it was forced in the mid-nineteenth century to modernize itself – at the time meaning 'Westernize' – to avoid colonization by the Great Powers. Thus postcolonial issues may pertain to questions in the study of cultures and thought of this period in which Japan was both a leading nation on the world stage and, like many aspiring empires, recently modernized. To Westernize was not necessarily a self-colonization, but a more neutral modernization, a translation of technologies (social as well as of the hard sciences) into local contexts in which the existing infrastructures (including language itself) became means enabling modernization and nation-building (unification) to preserve the polity in some contested form. The age of modernization, industrialization and capitalization is synonymous with nation-building and a spectrum of colonial activities. To refuse modernization would have been to invite colonization. To survive in this competitive international context, national unity (imposed through education and compelling narratives to mobilized people from disparate places, for instance) and mastery of the instrumentalization of human and natural resources at home and abroad dictated which geographical and demographical nations would thrive and what their place was among the dynamic global hierarchy. Japan was the first Asian country to modernize quickly enough not only to stave off colonization but to challenge and fight its way to be respected according to the rules the Great Powers themselves had used to dominate and subjugate colonies around the world. The known world had been transformed into a world reaping capital essential to nation-building.

In inventing an authentic philosophical engagement, someone born in Japan at the precipice of this potential national disaster would need to translate and comprehend Western philosophy as either an exotic discourse of others, a discourse recognizably evident in existing writing in Japanese, or something in between. After all, Japanese thought flourished in debates and publications in what is called early modern Japan, from the Tokugawa period (1600–1867), though with very little Western philosophy figuring into it during this period when Japan embarked on a 'closed country' policy in the interest of social stability after the Spanish use of Christian missionaries as a colonizing force was exposed. The *shōgun* (将軍 generalissimo) at the time had ruled Tokugawa Japan through an intricate system of controls over domains and relied in part on the limiting of certain technologies (especially weaponry) and Western learning. For European traders, only a small Dutch island port called Dejima was permitted, through which *Rangaku* (Dutch Learning) entered Japan and was introduced in Japanese woodblock-print books. Thus by way of covert Western studies, Japanese thought and science was quickly able to modernize when the new Meiji government, reorganized around the restored emperor, opened the country and encouraged rather than suppressed the study of foreign ideas already glimpsed through *Rangaku*. Since a trickle of Western (especially Dutch) books found their way into Japan before it embraced modernization, Japanese had devised translations of scientific terms using Chinese characters before the Chinese themselves, so that much of basic modern Chinese is derived from Japanese translations of modern terms originally in English, Dutch, French or German.[13] General issues of the day included, notably, what role an emperor should play, what responsibility the people (*min*) and the administrators (*kan*) should hold with respect to each other, and what role language itself should play (for instance the question of whether Japanese should continue to use Chinese for serious scholarship, Japanese or even French).

In any case, Nishida chose not simply to approach Western philosophy as a master discourse to be translated passively and allowed to exist as an exotic authority of others. Much as pre-Meiji Confucian-oriented scholarship localized Chinese scholarship so that it would diverge from mere annotations of texts and become a discursive means for deciding the fate of Japan, Nishida embraced the challenge of engaging Western philosophy on his own terms.

In the chapters that follow, the contested nature of nature itself, of materiality as an object of hard and social sciences and of the humanism (and post-humanism) of philosophical inquiry all present a portrait of an unsettling time in physics as well as global history in the last throes of colonialist expansionism. What we learn from this exercise in reframing the Kyoto School in light of the philosophy of science is that Nishida, Tanabe and Tosaka all have their enabling foci that can be more precisely understood in terms of how they each engaged questions concerning the undoing of assumptions supported by classical physics, as modern physics continued to demythologize space-time, causality and matter itself. This book shows not only how important physics is to the Kyoto School as a world philosophy, but also how through the study of world philosophies in general we enter by way of modern physics a world that is both post-scientific – in the sense that it demands humanistic supplements to what were once called objective occurrences – and transformative of these cultures. Physics is no longer a discourse on the given but a discourse on practices that have deconstructed the givens of space and time. Thus, though we have grown accustomed to physics offering as science a complete account of the physical world, it now forces us to reconsider our ontological and ideological predispositions in light of the changes it now imposes (not to mention technological innovations and cosmological changes in our understanding of the universe). Its demands are by no means to be misunderstood as forcing some fantasy of scientific realism on the world; such an assumption is the quintessential mark of not appreciating modern physics. Rather, by seeing what the Kyoto School philosophers have accomplished (even in some signs of refusing to understand certain points or exaggerating others) we can both realize that these brave path-breakers show us just how important physics is *even for non-physicists* and that by taking this path we can see them in a fresh light in comparing just how they approached these issues raised by relativity and quantum mechanics.

Scope of engagement with issues in the philosophy of science

As prominent writers with formal and informal followers, these three philosophers each at times serve as critics, producing various short commentaries. However, all of them have exhibited extreme interest in science both very early and very late in their adult lives and integrated it into their major philosophical works. None of them chose to go into science, and none of them appears to understand with the precision of a contemporary physicist (in Japan or anywhere) core equations expressing complicated relationships in physics, nor engage the problems of matrix mechanics in any detail. However, they each understood some particular aspect of contemporary physics in some detail, and related their own considerations of issues in physics to issues in their own emerging philosophical inquiries – not incidentally, but centrally shaping them in diverse ways. Nishida and Tanabe each resituated the core impulse of exploring an Asian void-orientated ontology in terms of Cantor's set theory (see Box 4.3), and both discussed problems of quantum physics in light of Bridgman's Operationalist critique (see Box 2.2) (though Tanabe's treatment of Bridgman is less significant to his oeuvre and is mostly omitted). Moreover, Tanabe understood one major physicist – Paul Dirac – in some detail, exploring the implications of a yin-and-yang of potential energy and 'void-holes' as yet another ground for nothingness, a recurring central concept in Kyoto School formulations of positions.

Aims

This book aims to provide access to Kyoto School philosophy for undergraduates and graduate students in any institution of learning offering courses in philosophy or the scientific humanities – a quickly growing area crossing interdisciplinary boundaries. It is designed to introduce issues as open topics for discussion, not reducible to pat answers, and therefore is well suited to lecture-discussion and seminars on the impact of science within history, political science, area studies (Japan, East Asia), comparative literature and various courses offered in humanities and social sciences programmes. As most of the translations are appearing here for the first time, the book may also serve as a resource for scholars and anyone interested in the philosophy of science, posthumanism, and world philosophies.

References

Friedman, Michael, 'Newton and Kant on Absolute Space: From Theology to Transcendental Philosophy', in Michel Bitbol, Pierre Kerszberg and Jean Petitot (eds), *Constituting Objectivity: Transcendental Perspectives on Modern Physics*, 35–50. Dordrecht: Springer, 2009.

Heidegger, Martin, *Being and Time*, New York: HarperPerennial/Modern Thought, 2008.

Heisig, James W., *Much Ado about Nothingness: Essays on Nishida and Tanabe*. Nagoya, Japan: Nanzan Institute for Religion and Culture, 2015.

Krummel, John W. 'Chōra in Heidegger and Nishida', *Studia Phænomenologica* 16 (2016): 489–518.

Levinas, Emmanuel, *Totality and Infinity: An Essay on Exteriority*, Pittsburgh: Duquesne University Press, 1969.

Maraldo, John C., 'The War over the Kyoto School', *Monumenta Nipponica* 61, no. 3 (2006): 375–406, www.jstor.org/stable/25066448.

Naoki, Sakai, 'Subject and Substratum: On Japanese Imperial Nationalism', *Cultural Studies* 14, no. 3–4 (2000): 462–530, DOI: 10.1080/09502380050130428.

Rigsby, Curtis A., 'Nishida on Heidegger', *Continental Philosophy Review* 42 (2010): 511–53.

Simao, Eikoh, 'Some Aspects of Japanese Science, 1868–1945', *Annals of Science* 46 (1989): 69–91.

Tani, Toru, 'Inquiry into the I, Disclosedness, and Self-consciousness: Husserl, Heidegger, Nishida', *Continental Philosophy Review* 31 (1998): 239–53.

Tremblay, Jacynthe, 'L'influence du concept de complémentarité dans la philosophie du dernier Nishida', *European Journal of Japanese Philosophy* 3 (2018): 57–77.

2

Nishida Philosophy, Place, Field and Quantum Phenomena

Introduction

Nishida's most famous work by any measure is his early work *An Inquiry into the Good* (*Zen no kenkyū*, 1911), a study focused on the subject to a degree he felt compelled to disavow later as he soon turned to historical and objective and dialectical renderings of subject–object relationality. His most enduring concept was *basho* (often translated as 'place', but also as 'field', chōra or 'matrix'), which shifts attention to objective presentation of being and the self not as a solipsistic point of access per se but rather as a co-productive rendering of self and being within an existing being that always begins with an undifferentiated void, or nothing, rather than *a priori* being. *Basho* is the site of this matrix of self and place, not simply place. His later work may be understood as a more abstract development of *basho* as a nexus of mutually and dialectically formative relationality based on sites of action rather than essences. The philosophical creativity Nishida exhibits in this work is not only a tribute to free thinking in general but a product of a talented, hard-working individual who thrived in his youth in an environment not hampered by excessive constraints of nationalist doctrine imposed after the promulgation of the Imperial Rescript on Education (*Kyōiku ni kansuru chokugo* 教育ニ関スル勅語, 1890).[1]

2.1 Nishida's method and the physical site of active intuition

Sensation and will in Nishida's pivotal turn from pure experience to poiesis, place and Minkowskian world-lines

Nishida's 'What Lies Behind Physical Phenomena' (*Butsuri-genshō no haigo ni aru mono*, 1924), included in *From the Acting to the Seeing* (1927), demonstrates how Nishida's concerns over physicality lead him to engage such aesthetic issues raised by Kant and others in light of contemporary issues in modern physics. He explicitly specifies the aptness of an analogy with 'aesthetic intuition' in describing the 'active self' and situating 'will' in this chapter. Here he presents a dynamic, interactive relationality of force, physical objects, space-time, will and certainty that invokes Minkowski's 'world-lines' in four-dimensional space-time.[2] He engages scientists such as Heinrich Hertz (1857–94), Rudolf Lotze (1817–81) and Herman Minkowski (1864–1909), as well as philosophers, including Johann Gottlieb Fichte (1762–1814), Franz Brentano (1838–1917), Immanuel Kant (1724–1804), Maine de Biran (1766–1824), Plato, Plotinus and unnamed others alluded to, including Albert Einstein (1879–1955), Georg Wilhelm Friedrich Hegel (1770–1831), William James (1842–1910), Friedrich Nietzsche (1844–1900) and Arthur Schopenhauer (1788–1860). The focus of this chapter is to trace Nishida's situating of will and certainty – which play key roles in how he situates human perception, the scientific method and the reflexive relationality of art that in shaping Kant's later thought on sensation and aesthetics will be shown to influence Nishida's thought and method. This minor work is overshadowed by slightly later work that would present his most well-known concept of the *basho* (site, field, matrix) but as such it may contribute to a new awareness of how we understand the formation of his thought. Nishida develops an argument situating the logic of the site for the human engaging one's will in acting itself so that the world and self both co-emerge in a joint making (*poiesis*); here, scientific and aesthetic engagement become two sides of the same coin in how the constitutive elements of a process of approaching the physical may be situated.

This chapter introduces many components of Nishida's argument, which itself is prone to integrating others' work to build his position. Each idea he borrows is carefully situated to show the role(s) it plays in the formulation of both his emerging system and his style. His style at the time reinforces his interest in a quasi-dialectical relation of subject and object that is original to Nishida, though incorporating elements of others. The range of philosophers introduced is broad, forming a periphery in support of his argument, which subsumes or transcends them. Each philosopher reflects established positions selectively brought into conversation within Nishida's modelling. His aim is apparently to focus on two poles not usually drawn together: the physical and the creative will. He aims to show how both scientific and aesthetic activity – as *poiesis* (Greek for 'making') – become frames of reference for developing an active site of subject–object dynamics that lay the ground for his subsequent original concept, that of the *basho*. Science and art both form adequate models for understanding both expression and presentation of the physical and the necessity of the human will in the process of framing both the physical in science and works of art. Nishida distinguished levels of awareness in ways that may at first seem counterintuitive, for instance, when we encounter the use of expression (*hyōgen*表現) as applied to the world via the human. Nishida's argument may also be seen to objectify the processes while calling for considered action that recognizes a complicated embodiment of an expressive subject in a play of relations that can be objectively described as based on fact-acts. As John Maraldo suggests, Nishida's approach to self and world offers new avenues of thought for studying ecological issues.[3]

As Masakatsu Fujita cites from Nishida:

> The idea of 'place' was concretized as the 'dialectical universal', and the standpoint of the 'dialectical universal' was made immediate as the standpoint of 'active intuition'. What I referred to as the world of direct experience or the world of pure experience [in an earlier edition of *Zen no kenkyū*] I now think of as the world of historical reality. It is indeed the world of active intuition, the world of poiesis, that is the true world of pure experience.[4]

Though in terms of Nishida philosophy the role of *basho* is central, if we wish to understand its origins, it behoves us to examine the 'What Lies Behind Physical Phenomena' chapter in his pivotal volume, *From the Acting to the Seeing*. Here we can begin to reflect on criticism of Nishida, documented for instance by Matteo Cestari, who highlights Tanabe's charge that active-intuition is simply 'an aesthetic and artistic action', an 'intuitionistic metaphysics, not far from Plotinus', known for an aestheticism that prizes contemplation over 'action and work'. Cestari also introduces Kōsaka Masaaki's accusations of contemplativism and aestheticizing *praxis*, which echoes a Marxist demand for materialist accounting of practices.[5] Even beyond the influence of materialism in this period, and later the danger of even mentioning Marx, 'What Lies Behind Physical Phenomena' offers a glimpse of an early attempt to justify human agency as more than either Bergsonian vitalism emanating from the world or Jamesian psychology.

Nishida variously argues for will to play a role in the very generation of space and time. While Kant saw space and time as products of *a priori* categories that enable perception itself and Hegel situated space and time as known by way of abstract intuitions, Nishida builds on contemporary Einsteinian relativity and Minkowski's world-lines in terms of the emphasis on differing subject (coordinate) systems that constitute space and time necessarily as space-time. Nishida's argument is fairly dense, integrating multiple sources to place *will* as central to co-emergence in the *poiesis* (making) of the world (which in itself sounds vaguely Schopenhauerian, but is not an argument primarily based on classical conceptions of force). What is being presented is not simply the will in itself, but both space-time and things as presented by the will. He then asserts that it is *action* of some kind that must trigger the presentation of space-time: 'Even though time and space exist, they cannot be given consideration without the acting.' But there is more than a passing conflation here; as Nishida emphasized in his conclusion to the first section of the essay, 'even when a change of position occurs in space, following Lotze, we can consider it a change in something's state [*jōtai*; or condition]. It is not that there are things in accordance with space, but that within things we can also consider there to be space.'[6] The point in introducing Lotze then is to bring the epistemological question of human knowledge and representation of space-time and things closer to an ontological presentation of them. This, indeed, reflects the endless reflexivity introduced by relativity theory and intensified greatly by quantum physics (not alluded to in this essay).

Then Nishida attempts to preserve qualities in the recognition of objects while also situating the human subject as interacting so as to somehow constitute the world: 'Without doing away with the characteristics of sensory qualities, if we consider mutual relations of the acting, something resembling time and space appears.' Thus, his sense of 'universal' in 'a universal [*ippanteki*] form of the expression[7] [*hyōgen*] of will' carries a sense of the universal as the commonplace, the shared community space-time as well as the individual will. It is not merely one will (either of the community or the self) but a pluralism that will be elaborated on in his later work *The Empirical Sciences* (1939) and elsewhere.

There is an implied complexity of shifting awareness here already, a responsibility shouldered by inner perception (or cognition), and a dynamic modelling predicated on temporally located subjects that access space and concepts in ways that *without* a site (*basho, chōra*, matrix, field, place) still imply a dialectical dynamic. One difference is that Nishida's focus here is on the physical, and this section underscores not aesthetics but rather the language of force in *classical* physics while tracking its emergence through Einstein and Minkowski into relativistic coordinate **systems** (see Box 2.1) that relate to one another without one self becoming ascendant (solipsistic). (Nishida's later works will address how this pluralistic world presents itself.) 'In terms of Minkowski's "world line," both [time and space] must always be incorporated as components ...

Box 2.1: Systems

In Newtonian physics the crucial assumption of a universal absolute space and absolute time that are continuous prevails. One could use Newton's laws to define the relations of motions of objects based on energy, momentum, mass, gravity and other basic factors in absolute terms. However, relativity debunked the absoluteness of space and time: they are bound up in one another and only accessible in local space-time systems, not absolutely. Then, quantum mechanics emerged out of nineteenth-century studies of electromagnetism and especially Planck's discovery in 1900 of the quantum of energy as part of the new field of studying radioactive particles. For modern physics, system indicates local points of reference that may include one or more particles or waves in relation to the observer and perhaps other bodies. In quantum mechanics, the particle positions in such systems are known only probabilistically and in the virtual realm (depending on the frame applied).

Physical phenomena are a relationship between one physical space and another. (In this way we may in terms of physics today begin to consider proximate activities.)'⁸ Thus Nishida prepares a ground for intersubjectivity evident in his later work on empirical science that is based on a potentially impersonal interrelation of subjects and objects in space, but only known through time as produced by human experience. The emphasis on 'proximate activities', rather than uniform Euclidean space evident in an aether-filled universe, also demonstrates Nishida's consideration of the importance of multiple subjects in relation to one another and the problems of defining objects both at sites and conceptually (in light of matrices, *chōra*) and fields of co-relation.

Many interpretations of Nishida have emphasized his debt to Zen Buddhism as indirectly present in his philosophy. This emphasis on observing the quintessential *will* of the self as objectively necessary may be recognized to be a form of relationality ostensibly grounded in continental philosophy, and yet it also invokes related concerns in Buddhism over how to situate desire in the self by emptying the self or seeing through the fictions such as the self that structures what we will. Below, when invoking Plotinus, the argument noticeably attempts to preserve human will against the reduction of it to an atomistic view of the universe as ultimately knowable apart from human will. This question will become even more important to understanding physics in later work reflecting the ongoing progress in quantum physics, where human interaction is integral to experiments.

Translation

What Lies Behind Physical Phenomena (*Butsuri genshō no haigo ni aru mono* 物理現象の背後にあるもの, 1924)⁹
Nishida Kitarō

1. The physics of proximate agency

By introducing 'proximate agency', Nishida begins to show how modern physics helps explain how our existential concerns over the origins of being out of nothing may be formulated not in terms of objects in space but rather the co-constitution of objects and space as fields of force. Things do not simply act; fields of force include object changes. Then he attempts to situate the observer in relation to qualities as continuous with change in space and time. Nishida asks how observers relate to a site (*basho*), a very important step in resituating subject and object in the fabric of local space-time.

In physics, one thing acts upon another. But in what sense can one thing become the cause for physical change in another? According to the physics of remote agency, one may consider one thing as immediately moving another, with the latter moving according to the force of the former. Thus one transcendent object acts upon another. However, according to the physics of proximate agency, to the very extent we adhere to such a linking of things to force we then find ourselves explaining physical phenomena by considering space a 'field [or site] of force'. Things do not act upon each other; it is according to the presence of things that space becomes a field of force. Thus, the phenomena of various things can be explained in light of change in the field of force. In this approach, physical reality shifts from the thing that acts to the space where the act appears, a space that seems to have physical qualities. But, what is it that provides the physical qualities in space? It is not that physical phenomena break away from the transformation of sensory qualities. Even when a so-called field of force is determined, it must take into account changes in the sensory qualities of all things. For instance, even in measuring an electric field, everything depends on various movements; but how can we make sense of such things moving? Even if some thing changes its position, we are not at the very site [*basho sonomono*] to see it. Given a coloured space with a sensory quality, something with a sensory quality can change its relative position; moreover, it remains merely a coloured space with a sensory quality. If a space having a sensory quality has no change in its relative position with another thing, [but] its sensory quality [itself] changes, we consider that not to be a matter of something moving but rather that something has changed temporally. However, even when change of position occurs in space, following Lotze, we can consider it a change in something's state [*jōtai*; or condition]. It is not that there are things in accordance with space, but that within things we can also consider there to be space.

2. Self-realization of will as basis for actual cognition[10]

By focusing on sensory observations, Nishida now further defines movement itself as predicated on the assumptions that qualities can form universals demonstrating changes in space and time. Nishida takes the implicitly Buddhist assumption of the interpenetrations of all things (including humans) as a point of reference that allows him to resituate the interdependencies of objects with the will of the observing subject. They are interrelated rather than simple givens as starting points of subject and object. Nishida thus challenges both the given of a person and the given of an object; the Kantian unknowable thing itself is brought to light not by way of abstract categories, but rather by an act of will that itself forms the *a priori*. Nishida thus is employing modern physics to inhibit a mechanistic view of the world, indeed a Newtonian worldview that underpinned Kantian phenomenology. When Nishida writes that act and knowledge are one, he is in effect situating the onus of movement on the human in order to verify being in a sort of perspective, which is built in this essay on points in time. The idea of space and time as a category builds on Kantian transcendental categories, but Nishida coordinates them in a more Hegelian dialectic that includes the category of the human will. Together, they verify what actually exists only in the present for the observing thinker whose will is at work.

Then, in order to understand movement, we must begin with the distinction of sensory content. For us to distinguish one [form of] sensory content from another there must be some thing that contains both of them; to distinguish red and blue, there must be something called colour universality [*iro-ippan*].[11] If we understand the experience of colour to be unable either to exist independently or in accord with something else, then we must see the concept of colour as itself containing an infinite potential for development. Rather than being treated as a simple abstract concept, it should be recognized as an idea [*rinen*].[12] In considering our visual sensory activity, it should be referred to as the consciousness of this sort of idea itself. Having now made colour into an independent experience, it follows that along the same lines sound too must be rendered an independent experience. Each of these independent things are mutually correlative and cannot be unified. We synthesize these various sensory activities in maintaining thought that unifies them.

Thought represents [*daihyō* 代表] sensory contents and is a higher-order activity in the sense of its synthetic unifying. Thought's representation of sensory contents is unified as action applied to discrete independent actions [*kakuji dokuritsu-naru sayō wo sayō toshite tōitsu-suru*]. Each mutually independent monad must in its reciprocal representation be contingent upon a higher-order monadic god-given pre-established harmony.[13] Speaking from the perspective of sensation [*kankaku*], the contents of thought may be considered to be nothing [*mu*]; however, insofar as sensory activity includes the world of colour and light, thought includes a world of thought itself, a creative activity. The object world of pure creative thought may be thought of as a world of numbers. However, a world of numbers is also a world of possibilities, not an actual world. The actual world, a world of forces, is not formed according to a simple *a priori* of thought. For Kant, the I in 'I think' is necessarily bound up in all representation [*hyōshō* 表象] and cannot configure an actual world, only at best a universally valid one in terms of simple logic. Kant seems to emphasize that in the very way we think our so-called experiential world is formed according to how sensory contents are combined. If the I in 'I think' transcends the individual person as a normative consciousness, there seems no way to form a single objective world combining reason and the contents of experience. Seen from [the perspective of] thought, this world of experience is a contingent world; however, seen from the perspective of the world of this reality, the world of thought is no more than a world of mere possibility. Even to say the experiential world is contingent in contrast to thought is not something simply made up. But, that phosphorus melts at 44 degrees Celsius is an unshakable fact. The laws of physics are laws according to these sort of facts. Though one might say that such laws attain objectivity simply by following forms of thought [*shii* 思惟], the objectivity of physical laws is not established simply by way of thought. One may say the contents of experience are particular in contrast with the contents of pure thought. However, it is by way of particular contents of experience holding universal contents of thought in check that the objective experiential world comes into existence. For such an object world to be formed, there must be an *a priori* by which the particulars of its ground are included within a universal, and our will is what contains the universal within the particular. Thus one can say that the so-called experiential world is formed according to the *a priori* of our wills. Without self-awareness [*jikaku* 自覚] of will, consciousness as a basis for the objectivity of knowledge cannot form. By way of self-awareness, act and knowledge are one. This unity is self-awareness. As a form of unification of such pure activity, the I [*watashi*] is an actual given category. 'Time' too must be a form of such unification thought of as a fundamental category of actual existence [*jitsuzai*]. Thus herein lies the distinction of one-dimensional continuity and 'time' as a category of reality. Time forms the self-awareness of the act itself [*sayō jishin*] while continuity is no more than its reflection. Taking the role of a point on a curve as an example, one may consider its continuity; however, indeed it is not something independent and unique; it is temporary, not 'time' that cannot be repeated. Objectified things are not equivalent to actual existence. The category of existence is not simply a category of thought, but rather thought is formed according to a synthesis of a self acting within this space and time. Therein, by way of the initial synthesis of pure apperception in 'I think', the realm of actual existence [*jitsuzai-kai*] can be formed. The true self must be something individual [*koseiteki*]. An I that is not somebody's I is not an I; consequently, no one can say what such an I thinks. If we can accept thinking along the lines of the above, even the category of a reality that forms a so-called experiential world comes into existence by way of consciousness of pure will. The simple contents of thought become objective knowledge according to a combination of space and time that holds the key to [*yūsuru*] sensory contents. When the will presents the self against the background of sensation, indeed consciousness first emerges. Ordinarily, since we think of sensation as simply passive, there is no determination of any objectivity; but, if one considers each sensation as an action, the unity of will that unifies pure action must incorporate a unity of sensations. As an aspect of will, pure apperception can unify sense experience. The objective will is a rationalization of the irrational, and as such gives rise to the objectivity of knowledge. To consider will to be unrelated to objective knowledge would thus be to consider subjective will alone.

3. The transformation of things

Recognition of change demands that the observing subject enter into a relationship with an object, which includes placing the subject in relation to itself. External causality is discounted as not necessitating internal change to an object (whether or not animate). Change entails having observed an emergence.

Since we cannot intuit time and space in themselves directly, we rely on the movement of things, and therein on a change in quality. To discern a change in quality, there must be a judgement of a distinction [*shikibetsu*]. However, that called change cannot be known only by way of a distinction. We can consider a thing's various qualities, that is, consider which propositional subject regarding the qualitative judgement of a thing will be entertained. In this case, although various judgements are bound together by way of the [chosen] propositional subject, still the change of a given thing eludes us, and we are simply left with a static unity. When just two mutually independent things enter into a mutual relationship in some sense, that is, if in some sense they are unified, this must be attributed to mutual change. In the case of just one thing, change eludes us. Of course, I suppose there are also cases in which the cause for a change is internal to something, as in the case of living things. However, in considering living things as independent bodies not in the same sense as something formed by elements, they may be thought of as a [form of] unity of an even higher order. Still, to see living things as changing by themselves is to see unity itself as equivalent with its elements. Things cannot move by themselves. From a certain standpoint, for an individual to be thought to move there must necessarily be another individual to be considered from the same standpoint. For one thing to change, there must be some equivalent independent thing to serve as the cause of its change. If a thing completely loses its independence when it is changed by another [thing], then it a case of it never having been changed in the first place. Change in a thing can only be conferred by way of a standpoint observing it directly as one with another [thing] along with the formation of the thing independently. The very formation of this thing independently, in other words in terms of the standpoint by which it is seen as one with another thing, first enables formation of change in a thing. Physically, for instance, since a thing is formed in terms of space and time, change and movement in a thing can also be said to be bound up in space and time, and a thing can be said to continue sustaining the thing itself and moreover change it.

4. The formation of the concept of the individual

Nishida then explores changes in space-time in relation to a thing, and whether a given thing retains independence. This framing question itself demonstrates his interest in situating human will, an issue foremost in the work of William James, whom he often cited.[14] This section brings into intense conflict the competing needs for both a subject and an object in a world bound to the local yet aspiring to make absolute claims. Nishida argues that formal propositions in effect present willed judgments of objects and their activities and manifestations of qualities. Continuity can only be generated with respect to a human self. Quality is not of an object but of the perceiving human mind, and is articulated in propositions by which qualifying predicates may proliferate infinitely. While the will is crucial, solipsism is not an option for Nishida, since one becomes engaged in mutual relationality with individual things. What is recognized to exist, in effect, depends on our purpose.

Here we must address a question. Given that a thing formed in space-time changes, if the state of the space-time changes, should we consider the thing to have lost its independence as an individual thing [*kobutsu*]? Change in a thing can also be seen as a modality with spatial and temporal attributes. By contrast, if we distance ourselves from time and space and then attempt to consider some thing, we would be left with only uncertainty about such changes in it. How should we situate the independent individual thing here? To consider something an independent individual thing, we will first think of it as existing apart from our subjectivity and then think of how to maintain it in itself by realizing how it is bound up in various relations in the realm of objectivity [*kyakkan-kai*]. Bound up in an infinite number of relations yet maintaining itself in itself suggests we may consider it to possess an infinite quality [*seishitsu*] and at very least that it must take the form of a [logical] subject [*shugo*] of an unlimited number of predicates.[15] As a material property, even when taken to be finite, it must be considered an infinite subject [in a logical proposition] of a predicate in terms of its spatio-temporal relations as well as in terms of how it possesses the possibility for interacting with other things in an infinity of relations. If not the case, if merely taken to be a universal concept, it might be reduced to the subjective. Thereupon, in considering one individual thing, there above all must be a subject [*shugo*] [of logical propositions] of an unlimited [number of] judgements, and it must be what connects and unifies the judgements. Then, in proceeding to join and unify an unlimited number of judgements, a fixed direction is required.[16] When its direction is finite we can therein discern an independent individual thing. Regardless of how far this direction goes, just that it proceeds does not allow for the formation of a concept of an individual thing. To form the concept of an individual thing, there must be a transcendence of point of view. Of course, it would seem inconceivable that simply calling something infinite would then render it an independent individual thing [*kobutsu*]. No matter how infinite what should be done [*tōi*, cf. *Sollen*] is, it does not amount to an independent individual [*kotai*]. In the gap between object and action, the true concept of the individual is not established. In constituting a concept of the individual, action and object must be related. The truly objective must take the form of an object *qua* action or an action *qua* object. No matter how infinite things appearing before me and things appearing in me seem, they can still be seen as things within me.[17] That the self cannot be truly transcended by way of the self, however we may attempt it, means that there must be something appearing from behind to envelop the self within it. The continuity of limitless contents of experience on the one hand may be seen merely as our mental activity [*seishin sayō*]. Yet such activity, when objectified from the standpoint of the agency of activity [*sayō no sayō no tachiba*], becomes a force [*chikara*] transcending the self, so that the subject [*shutai*] endowed with such force may be thought of as an individual thing [*kobutsu*]. An individual thing forms as an object of the agency of activity. We can by taking singular things as the subject [in a logical proposition][18] proceed ad infinitum to add any number of qualitative predicates. However, we cannot take our actions in themselves as the subject [in a logical proposition] and match them with these predicates, and we cannot say that the red in the activity of our perception is red [in itself]. But when these activities in themselves are objectified, we can treat the colour as an activity of a thing. Thus to objectify it must not simply entail an intellectual self-awareness [*jikaku*] of unified judgement, but rather a willed consciousness. Even though we can initially seek in intellectual self-awareness this limitless inner unity of judgements, it alone will not encompass judgements concerning the workings of things, only entertain the continuity of judgement in terms of oneself.[19] When attaining consciousness of an inner unity that gives rise to such continuity, that is, of a universal in itself that gives rise to such a judgement, quality is not something external to us, and judgement and its contents cannot be distinguished, but rather contents are the basis for judgement and predicates integrate subjects.[20] Colour and the judgement of colour cannot be disentangled. What objectively becomes a judgement of colour must be colour itself, since a judgement of colour forms as the development of differentiations of the colour universal. Judging consciousness [*handanteki ishiki*] cannot proceed as such; moreover, when we see what forms the basis [for this differentiation], it may be considered to be more than judging consciousness, something resembling the so-called limitations of cognition [*ninshiki*]. However, though these limitations are transcendental, it is not exterior to the self. When we consider activities of the self initially as such activities in this way, and seek the basis of subjective consciousness in itself, these various qualities, in other words, may be considered secondary qualities of things presented as one's sensations. Such a ground for consideration depends upon our selves' unity of independent activity. Accompanying the independence of each activity in our consciousness is a synthesis into one. However, by contrast, just as visual sensory activity can be understood as neither red nor blue, a unity in itself while encompassing all its activity necessarily homogenizes its contents [*izure demo nai*]. When such unities are objectified, the concepts of individual things are established. What does it mean for something to be objectified? Transcending this unity, there is something nevertheless containing this something within itself. Thus, transcending the mental consciousness of the interior unity of unlimited activities, the will must serve as that which contains [this objectified unity] within itself. From the standpoint of will transcending the conscious self, all things that move are to be seen as things that change. Due to the transcendence of the thinking self, all things are each independent individual things; and yet, in their mutual relationality, they enter a single unity.[21] Since each personality is free, it is common to construct a single 'kingdom of purposes'.

5. The inseparability of time and space as categories of existence: time and space as expressions of will

The title itself states Nishida's argument, which forms the key model developed implicitly in the remaining sections. In short, Nishida argues that Minkowski's four-dimensional space-time (three spatial dimensions of x, y, and z with a timeline t) provides a basis for situating human will. He identifies the Minkowskian world-line of time as the moving site of the individual. Recall that it was Einstein's use of the speed of light (c) as the invariant forming the possibility for the co-variants of time and space to form a point of emergence of common space-time itself visible from multiple perspectives (relativity does not mean open to any positioning, but rather that positioning itself must be recognized as precise although dependent on local observers). Such co-variants as time and space made the theory of relativity possible and necessitated the rethinking of the classical relations of causality in relation to energy rather than mass and velocity ($E=mc^2$). Nishida in effect offers a philosophical exploration of how subject–object relationality may be reinterpreted in light of the emergence of a co-variant space-time structurality in relativity. At this point, Nishida focuses on how relativity itself divorced classical physics from modern problems that treat certain questions in term of time (waves) and others in terms of space (particles or Newton's long-rejected Corpuscular Theory of Light). Light becomes a common nexus. The local site in both cases still requires an observer, as in all science. Nishida situates the will explicitly as necessary in modern scientific thought and in general. Subsequent sections situate will variously in light of interior cognition and observable nature, and though Minkowski and Einstein are hardly mentioned, their thought forms the groundwork for this essay.

Although space and time are usually considered to be a form of mutual relations of the acting [*hataraku mono*], I think one should regard them rather as a universal [*ippanteki*] form of the expression [*hyōgen*] of will. Even though time and space exist, they cannot be given consideration without the acting. More than just things acting in time and space, as Lotze said, within a thing there is space and time.[22] Without doing away with the characteristics of sensorial qualities, if we consider mutual relations of the acting, something resembling time and space appears; moreover, it is in the absence of curved space thought to be equivalent with Euclidean space. Though usually force can be considered in light of a synthesis of space-time, I on the contrary see force first of all as an expression of our will,[23] while space-time may be thought of as its universal form. As space and time exist together as a form of so-called actual existence [*jitsuzai*] they are not separate, being necessarily synthesized internally. Without sensory engagement [*lit.* resistance or opposition, *teikō*] there is no actual space,[24] as it is in engagement [entailing resistance] that the consciousness of force is required. Activity must incorporate time. Even if space is thought to be something still, stillness cannot be separated from time, as something still is stilled in terms of time. Time, moreover, cannot be considered something devoid of activity. Though activities are thought to be acted out in time, without the acting out, actual time cannot be thought, and time as a void becomes nothing but mere thought. In terms of the physical world, time is always synthesized with space.

Physically concurrent with time, that called the passage of time always incorporates spatial relations. But, the phenomenon of consciousness is merely temporal, and believed to be unrelated to space, since activity is conceived in terms of time. However, even though called a mental phenomenon, it must somehow synthesize[25] an objective space in light of birth and death[26] in terms of objective time. We do not consider merely mental time as true time. Even though such 'time' is incorporated within a conscious phenomenon, a conscious phenomenon does not occur at such a 'time'. Though time is said to be one-dimensional, the 'time' in which conscious phenomena are born and die in it does not resemble merely a one-dimensional straight line; yet, to say we can consider this spatial region to be zero would be equivalent to seeing time as zero within a motionless space. In terms of Minkowski's 'world-line', both must always be incorporated as components. In a world devoid of time and space, matter must be such that it can be shifted in parallel (*Parallelverschiebung*) uniformly. Actual space and actual time must always be physical. If we approach them in this way, the independence of individuals, problematized above, would now seem to lose its sense of forming a contradiction out of the object's loss of its independence in relation to space-time. Things do not move in empty space and time; the empty time and space that can be considered separate time and space are nothing but abstract concepts.[27] Physical phenomena are a relationship between one physical space and another. (In this way we may in terms of physics today begin to consider proximate activities.)

6. The mediation of force and will as inner perception

Here Nishida returns to interior questions of situating the subject in terms that we now can see to be an attempt to mediate between the assumptions of classical and modern physics. Nishida seems to suggest a parallel or symmetry of some kind between material forces and human will in relation to space-time, at least insofar as some sort of meta-critical self-awareness allows a bigger picture to be seen. Ultimately, he argues for the importance of will in ascertaining facts and object worlds. Here Fichte's fact-act is enlisted in effect to define the emergence of self and world in terms of impermanence, a cornerstone of Buddhist thought, in its confinement to the present. This limitation is backed in modern physics by Einstein and Minkowski, who Nishida has drawn upon to reach this fulcrum of interior and exterior, and by Brentano, who downplayed spatial continuity in favour of focusing on 'change in time'. Space, in Nishida's argument, becomes predicated entirely on inner perception. Then, self-awareness itself is argued to have its own limits as self-reflection opens to infinite introspection, which forces us to exert our will (for instance, to choose what to focus attention upon). However, Nishida closes this section with a suggestion that the use of vectors and tensors to define complex force-relations in the exterior world complicates the degree to which the descriptive interior syntheses may maintain efficacy. In fact, similar questions have continued to plague the philosophy of physics to this day, as it constitutes a line of inquiry into questions over how visualizable relations in relativity theories may be supplanted by non-visualizable numerical data in quantum mechanics. (See also Tanabe's discussion of Dirac in 3.2, and Chapter 5.)

In relating physical spaces one cannot assume there is a physical space. Even in considering physical force as introducing empty space, there is no empty space, but merely to the degree there appears a substance that fills it uniformly. Moreover, time is not a synthesis of space from outside, but it is rather incorporated in physical space. In that case, what constitutes the physical world in the mutual interrelation of physical spaces?[28] I think one must look for it in our self-awareness of will [*ishi no jikaku*]. Just as material force is expressed according to space-time but is not contained within it, will is expressed according to material force but is not contained within it. While without the expression of space-time there will be no material force, without the expression of material force there will be no will.[29] Our wills incorporate expression in a supra-conscious world [*chō-ishiki-kai*]. However, even though material force is known by way of coordinates in space-time, it is not contained within them; on the contrary, as if establishing it [that is, space-time], even though in attaining consciousness the will sees the world of force in an object world, the will is not contained within the world of force, but rather it is through the will that the world of force is formed. In order to clarify this point, I wish to dwell briefly on inner perception (*innere Wahrnehmung*). Actual knowledge must be grounded in inner perception.[30] Even in terms of an occasion when we recognize a fact of the outer world, inner perception must form its basis. The standpoint of consciousness offers objectivity in all factual knowledge. Even in accepting the suggestions of mathematical principles in a dream, there is no barrier to mathematical principles for us. However, a physical experiment carried out in a dream would carry little authority as a truth. Incorporating both the world of dreams and the world of existence in self-awareness [*jikaku*], we are in a position to be able to distinguish dream and reality. How then is factual objectivity produced according to self-awareness? In self-awareness, the end is included in the beginning, since consciousness of self-awareness is formed by a single development that recalls its origin. That called interior perception also knows itself in terms of such forms of self-awareness. However, self-awareness is not simply a closed circle, but rather like Fichte's fact-act [*jikō; Tathandlung*] an infinite procession.[31] Self-awareness of one's distinct self arises accordingly, and the form of 'time' is considered to be unrepeatable. From this standpoint, thought and sensation can be considered to be synthesized within us.[32] We perceive internally the contents of both thought and sensation. Against the standpoint of interior perception, the activities of thought and presentation both occur in the same place.[33] As Brentano argues, when seen from this standpoint, along with an idea [*hyōshō*, tr. of *Vorstellung*, tr. of idea] not being individual [*kojinteki*], concepts also must remain intuitive. Given the above, our visual sensory activity too, as interior activity of the self, must take the form of infinite development of both sensory contents and, in the realm of thought activities, an infinite development of thought contents. As the contents of consciousness, for thought to be capable of approaching any kind of sensory contents, it must accommodate it. As such, though factual knowledge is formed in conjunction with interior perception, there must be self-awareness of acting [*hataraku mono no jikaku*] as the basis for such interior perception, and the objectivity of interior perception is presented according the self-awareness of will. Though on the one hand we consider self-awareness of will to be impossible without interior perception, the truth is that without self-awareness of will there can be no interior perception. Our selves are not a series of phenomena in time but rather unities of activities themselves. From the standpoint of self-awareness of will, we see the positive [*sekkyokuteki*] contents of inner perception. Brentano posits that in considering the modes of presentation (*Modi des Vorstellens*) in the past, present and future, there cannot be presentation without a temporal mode (*Temporalmodus*) just as there cannot be judgement without a qualitative mode (*Qualitätsmodus*). This thought is based on his distinction between a direct viewing mode (*Modus rectus*) and an oblique one (*Modus obliquus*). When we hear a series of sounds, one sound appears as the first present, and one after the other they continue. However, since we know the self's consciousness of such phenomena in the present, thoughts of what has passed are also found in the present, that is, are presented obliquely.[34] Thus just as temporal relations form a subset of inner perception as a mode of consciousness, spatial relations must in the same way be included. According to Kraus,[35] Brentano considers change in time as primary and change in place as secondary; in other words, continuity is considered secondary. Without a relation to the innermost depths of perception, we cannot consider that called space. As a straight line extends endlessly both right and left, even if thought to meet at a point in infinity [*mugenten*], we, in short, in our inner perception make the point of the viewing mode [*chokushi yōsō*] the centre, and according to an oblique viewing mode proceeding back and forth without end;[36] but, does it not seem reasonable that we can see things objectified and contained in terms of the viewpoint of self-reflection on inner perception itself, in other words, in terms of the viewpoint of consciousness of activity itself? We might consider the thought of a straight line too to be formed in accordance with the inner perception's determination of the self within the self. Though inner perception itself is habitually considered to be temporal insofar as it exhibits inner continuity with acts, the inner perception that opens objectivity to knowledge of facts [*shijitsu no chishiki*] that is not usually considered simply temporal must include spatiality within it. In our consciousness of self-awareness in our activities, when we incorporate within us inner perceptions we transcend a world of force [*chikara no sekai*] that in its form incorporates both time and space. We think sensations have degrees of intensity, and that their intensity is found in standpoints of inner perception. However, entering the innermost depths of inner perception itself, when transcending the distinctiveness [*tokushusei*] of the contents of experience, the intensity of sensations become spatio-temporal as a modality of inner perception. In the principle of the 'anticipations of perception',[37] the idea is that sensations possess levels of intensity according to which they become real. Moreover, when transcending the position of inner perception and standing in the position of the self-awareness of will, one moves from the world of Kant's *mathematische Grundsätze* to the world of his *dynamische Grundsätze*,[38] in other words from the world of perception to the so-called world of experience. At the bottom of inner perception that may be considered infinite inner regress (*Regressus*) infinite outer progress (*Egressus*) must be included. In addition, as in seeing a form (*katachi*) as the determination of a formless form, one sees the act of self-reflection as a determination of an inability to engage in self-reflection. Where self-reflection has become difficult one will surely find a world of infinite activity. From this position the infinitely introspective self is able to form. Thus, the site of the formation of such an introspective self is always the present, and it is in this setting that our will functions. That called the present does not merely refer to temporality but can include spatial relations. That is, since we are cognizant of what is called force, inner perception might be considered something deep rather than the consciousness of force. In other words, we may consider ourselves capable of objectively viewing our inner perceptions. However, the formation of the concept of force cannot occur simply from the standpoint of descriptive inner perception. If we see a physical phenomenon strictly from such as standpoint, as is reflected in Hertz's dynamics [or mechanics; *rikigaku*], one might dispense with the word *riki* [or *chikara*, force] in *rikigaku* [(the field of) dynamics]. But, even if one removes the word 'force', there must be something which organizes the physical world and unifies time, space and mass. As the physical world possesses objectivity, what structures it must be *a priori*.[39] Our conceptual knowledge is always analytical. Even in coming to know physical phenomena, knowledge of them must depend upon distinguishing them in terms of time, space and mass. However, physical phenomena are not simply syntheses of these various things. There must be principles that structure the functional relations among these various things. I think that in the ideas of fourfold-force (*Viererkraft*)[40] and energy momentum tensors (*Energie-Impulstensor*) in the physics of relativity theory one finds an all-encompassing appearance of truth [*shinsō*] in physical phenomena, amply providing the principles of structuration for physical perception. It goes without saying that the acquisition of such structuring principles cannot be based simply on descriptive inner perception. Considering physical phenomena in terms of vectors (*Vektor*) may provide a degree of practical clarification, and I think it may already constitute a refinement of our knowledge of the physical.

Above, Nishida invokes 'energy momentum tensors', also known as 'stress-energy tensors', which describe energy and momentum in space-time according to density and flux. In effect, they integrate energy into classical physics' approaches to quantifying stress tensors. It may be helpful to understand stress tensors and stress-energy tensors as describing classical continuum mechanics based on a comprehension of matter as a continuous mass, yet doing so in ways adaptable to relativity theory, which seems to be Nishida's implication. Use of tensors in quantum mechanics is less compatible with classical mechanics, yet has applications (see Chapter 3) combined with probabilistic approaches and uses of matrices representable as tensors.

7. Certainty as the mediation of inner perception and consciousness of force: certainty and clarity

This section situates certainty in relation to will. Certainty indeed may be understood in the context of this essay as developing Einsteinian insights on relativity based on the invariant of the speed of light, such as when Nishida invokes Brentano's distinction of certainty and intensity of consciousness respectively as shared and individual modalities. Though passing over this argument very quickly, he seeks to frame a Cartesian clarity of the beholder as inconsequential with respect to intensity (it could be a delusion), and relies on certainty as 'a modality of consciousness of transindividual thought' in a way that agrees with commonality of the speed of light to determine multiple positions in a relativistic frame of reference. Nishida also reflects somewhat recent currents in German aesthetics, invoking the then well-known German psychologist Theodor Lipps (1851–1914) apparently to bolster Nishida's emphasis on merging the subject engaging the world with the object perceived. Lipps promoted an aesthetic view by which the subject projected the self onto the object through acts of empathy. Nishida was likely moved by passages (in the original German) such as this:

> For what I feel is, quite generally, life. And life is energy, inner exertion, striving, and achieving. Life can be summed up with one word: activity, freely owing or inhibited, easy or effortful, in harmony or in conflict with itself, tensing and relaxing, concentrated on a single point or dissolving into life's manifold actions and 'losing itself' in them.[41]

Nishida goes on to distinguish direct contact with things and one's ability to distinguish how we are also conscious of the process of determining evidence of our perceptions and intuitions. To support this distinction of direct and self-aware engagement with objects, Nishida turns to Maine de Biran's division of active and passive impressions, which carries implicit weight in light of Nishida's use of Fichte as well as Lipps, who both underscore the importance of action of the self that constitutes engagement with the world. This positioning is crucial to Nishida philosophy, and this essay demonstrates how relativity theory in physics provides a grounding metaphor for placing the subject on a continual timeline in a world of change observed by the subject in Nishida's modelling. De Biran provides a somewhat surprising element here; though one might expect Fichte and Lipps to lead Nishida to what might have amounted to a more Romantic creative subject, on the contrary what is proposed is more in keeping not only with relativity theory but with a Buddhist grounding in a given world of illusions that fall away to reveal the nothingness of existence that the self itself conjures within human conventions that are tangible givens yet neither objective nor simply abstract things. Though Nishida does not mention Buddhism, the movement of his argument itself accords with such a deconstruction of conventions. Thus, upon seeing the word 'convention', one might refrain from rushing to judge Nishida as supporting the status quo; he merely finds in conventions (as a Buddhist) a form of certainty in patterned familiar behaviour that cannot be reduced to objects per se (Buddhism postulates the self to be a fiction, the non-self to be the revelatory truth, and convention in this sense something illusory but part of how we construct social and natural meanings meant to be seen through). Nishida writes, 'Conventions are but the contents of the active self reflected in the world of recognized objects.' Following de Biran, Nishida retains in the example of manifestations of convention a distinct sense of the complexity of attaining certainty, and how engaging 'active sensation' provides in effect a deconstructive methodology not limited to obvious abstractions alone. Furthermore, at the close of this section, to affirm the necessary element of conscious will to take responsibility in the world of physical forces and positions, Nishida argues that it is only by way of such 'active projection in the world of consciousness' – as suggested by Fichte, Lipps, and de Biran – that the mediation of the laws of nature as well as human convention are each synthesized through agential engagement.

The phraseology below suggests a debt to Schopenhauer and Nietzsche's development of the relations of force to the construction of reality itself. What seems to be missing in Nishida scholarship is an understanding of how he accounts for the role of the 'degree of intensity' as physical and mental processes invoking elements of scientific method, phenomenology, modern critique and numerous canonical and contemporary philosophers.

In order to clarify the relationship between inner perception and the consciousness of force, I would in addition like to inject a consciousness of conviction (*Überzeugung*) between them. Inner perception and certainty form a relationship that should not be broken. While there is no pairing of inner perception with certainty, without inner perception there can be no certainty; thus, to form a consciousness of certainty depends upon a unification of subject and object. Through consciousness of certainty, inner perception transcends the self itself and sees the eternal object world. All the object world is built upon this consciousness of certainty. Even an individual's mental facts, depending on one's certainty about them, become truths that must be recognized by anyone. Not only do factual truths simply depend on this but even ostensible universal truths. Everyone is bound to considering certainty in this way. At this point, in transcending our so-called individual selves, each of these [truths] depends upon the inner perceptions held by an infinite number of selves. Though Brentano distinguishes certainty and intensity of consciousness, cases in which the intensity of consciousness is considered an attribute differentiated from quality of consciousness, as Lipps suggests, may be thought of as an imposition (*Zumutung*). Therefore, while one should distinguish it from certainty, the modality of its consciousness is synthesized internally. It seems to arise in relation to it just as illusion and delusion do. If one may consider intensity the modality of sensory consciousness, certainty may be said to be a modality of consciousness of transindividual thought. Here, not only should one distinguish the two but also combine at one point that called the self as a modality of consciousness. Of course, in terms of certainty, one might say we still truly transcend the self and directly come into contact with [*chokusetsu suru*] objective things. However, in terms of a consciousness of clear *Evidenz* [evidence], we indeed lose the self itself and can come into contact with objective things and intuit truths. Accordingly, the will loses the self itself. Although one may see the value of clear sentiment as a given, one cannot even consider value apart from clear sentiment. In other words, we may say that in terms of factual knowledge, one may not arrive at clear sentiment in the same way as with pure theoretical knowledge. However, even at the foundation of certainty in factual knowledge there must be some objective thing that transcends the so-called self. Maine de Biran recognized a fundamental distinction in the difference between passive and active impressions. While passive impressions may gradually disappear as common habits, active impressions will remain clear. For conventions, that is, for what is repeated, one maintains clear contents of consciousness. While it may not constitute so-called knowledge, our certainty attends to such contents of consciousness. For instance, given our familiarity with a technique, it is in accordance to our familiarity with it that we can entertain a specific kind of certainty. However, we can think of this feeling as in the end approaching a kind of clear feeling in attaining something resembling artistic agency. In it, we come in contact with some objective thing. As in encountering a truth that should not shift in terms of what we know about it, in losing the self itself we see some things as objective. While the contents of [a passive impression] may be seen as impersonal and neutralizing thought, the contents of [an active impression] may be seen as personal and neutralizing the will. In considering how the qualities of certainty differ according to whether based on clarity of sentiment or convention, note how in the case of a convention we do not directly see it objectively, so that the contents of [its] certainty must constitute a kind of intellectual contents. Yet I, like Maine de Biran, think that the contents of certainty attained in convention is a contents based on active sensation. Though certainty cannot be formed without some objective basis and things that are not true according to convention cannot be considered true, things without objective bases do not appear according to convention either. One can only know the contents of an act [*hataraku mono no naiyō*] by way of the fact of an act [*hataraku koto*]. In this sense, convention may be seen as one kind of intellectual agency [*chiteki sayō*] by which the self clarifies the contents of the self.[42] If we posit that memory preserves memory itself and develops the self itself, when we reach something truly objectively in this way, in other words when we attain subject–object unity through intuition, certainty changes into a kind of clear sentiment. Artistic intuition exhibits just such clear sentiment, yet when reaching a truth through thought the clear sentiment attained is the same. When one arrives at clear sentiment out of certainty, all that seems to be inner perception is erased. Inner perception is erased in the sense of objective things encompassing subjective ones and self transcending itself. Just as thought is considered to be supra-individual, from this standpoint, inner perception is transcended and inwardly unlimited inner perceptions are formed. In artistic intuition, space and time are not located outside but are included within it,[43] even as non-intensity [*hi-kyōdo*, 否強度], since colour also exhibits intensity in paintings. When based on inner perception, it gives rise to unshakable factual knowledge, that is, we attain certainty with regard to factual knowledge. Even if only as seen in a dream, being seen in a dream is a fact. However, at the bottom of such inner perception there must be a consciousness of self at work as well as a self-awareness of will; a world of force is formed based on them. When the self recognizes the activity of the self and the self engaged in an activity becomes the knowing self, in brief, the category of causality is formed based on the self-consciousness of the active self. In the will's realm of object-world forces [*ishi no taishō-kai-taru chikara no sekai*], what becomes a closed system must be the rules of nature. Here we become independent of coordinates. The coordinates used by physicists form a system of a non-existent time seen from the standpoint of inner perception.[44] As Fichte noted, the non-self opposes the self, while at its foundation there is absolute self; the opposition between self and non-self is built upon the absolute self. In this way, from the standpoint of the absolute self we preserve a clear subjective *qua* objective world, and because of the inclusion of the questioning self we preserve an object world beyond doubt.[45] In terms of the world of certainty, though the self is still not distinct from the conscious self, in terms of the clear world, the self stands completely apart from the conscious self, just as in the case of contemplating purely theoretical clarity. Taking the standpoint of the conscious self, such an object world constitutes an unreachable world of force [predicated on] endless activity. While an inner continuity borne of deep and endless reflection is formed in the conscious self, it cannot transcend this continuity. To do so would necessarily depend on the active subject. According to inner perception as elevated by such a standpoint, something resembling a site of force becomes visible, and all lines become lines of force. Of course, the physical world is not a perfectly clear world. However, Maine de Biran argues that for the voluntary effort (*l'effort volontaire*) to be made the basis of the general idea of causation there must be active projection in the world of consciousness. Conventions are but the contents of the active self reflected in the world of recognized objects.[46] What are clarified according to conventions are objective contents from the perspective of the active self. The laws of nature may be thought of as conventions of a transcendental self. By negating subjective conventions of the self, natural scientists clarify the objective conventions of nature. All of what appears by convention forms a single objective world, and the conscious self is submerged within it. Seen from the standpoint of the supra-individual conventions of the active self, the so-called world of nature appears. As only intellectual selves, we cannot know the conventions of nature itself from within, and for this reason knowledge in terms of the natural sciences is thought to lack perfect clarity. In terms of the standpoint of active self, attaining perfect clarity requires a synthesis of subjective and objective agencies.

8. Various laws of causality in the formation of conscious will: mental and physical forms of causality

While section 7 argued that 'the category of causality is formed based on the self-consciousness of the active self', here this causality is explored in more detail. The self is presented as an aesthetic self in a sense that integrates modern relativity physics – mentioned here – with philosophy. Physics is invoked in a modern sense that calls upon the observer situating in time and space, not in the classical sense of Newtonian causality. Whereas in Kant (following Newton), objects appeared according to transcendental categories in chains of phenomenal displacement situating subject and object (though never verifying the thing-in-itself), Nishida will in his work in the wake of this essay move closer to Hegelian dialectics situating subject–object relationality within a topos (*basho*)[47] outlined here as dependent on a space-time and a 'truly active self seeing the object world'. Moreover, this section formulates mental and physical phenomena as both constituents of the active self, which according to the withdrawal of will allows a natural world to emerge as physical phenomena. Conversely, when time is extended, the mental phenomena gain prominence. He then situates 'the *a priori* of the will' as another means of distinguishing these two types of phenomena. Invoking Plato and especially Plotinus, he then turns to creative will as the ultimate ideal for structuring the physical world. This theme of reflexive self-awareness that is also always engaged with the physical world is a theme that defines Nishida philosophy and is found in most subsequent major works.

The active self immediately returns to the self itself. What sees in the self the perfect world of objects[48] must be at the site of the subject–object union that truly sees the act in itself; in other words, it must be from a standpoint that resembles aesthetic intuition. From the standpoint of thought, no matter how one tries, the contents of the active self cannot be objectified, but only seen as something acting upon all things [*subete no mono wo ugoku mono toshite*]. Yet from the standpoint of the thinking self, since the innermost self cannot look back on itself, in terms of the object world, it is upon seeing each moving thing that thought resembling remote agency will arise. However, when the active self looks back inside the self itself, the self is active without being engaged in an activity other than the unification of agency. In thinking from such standpoints, one sees the object world as a field of forces, and thought resembling proximal agency arises. In the prior instance [of remote agency], what unifies sense contents is a still self, while in the latter case [of proximate agency] is it an active self. Putting aside time and space, what combines force to force is the subjectivity of the will [*ishi shukan*]. Force not only includes space in its expression of will but also necessarily time too. The active self does not operate within time; it includes time within it. Within the physics of relativity theory, space and time become inseparable from each other, and one can entertain the possibility of the truly active self seeing the object world. It is in terms of what contains time itself within itself that we can best see the appearance [*eizō*] of the active self. I think that one can thus consider what appears to be a relationship between mental phenomena and physical phenomena. Our mental phenomena too, while appearing temporally, incorporate time within them so that the contemplation of significance *qua* actual existence [*imi soku jitsuzai*] depends on it. The realms of both types of phenomena constitute the object world of the active self, together merely distinguishing two types of form. I believe the distinction of two types of phenomenal worlds is based on the contents of time. What I mean is that the natural world arises from a standpoint of one's negation of will [*ishi-hitei*]; the physical world arises when temporal coordinates are formal and rendered devoid of contents. For physical phenomena, the shortest distance [literally 'line'] may even be considered in the object world of an active self. On the contrary, when time possesses positive contents mental phenomena may form. What does it mean to have positive contents of will? Though in terms of knowledge we consider the subjective to follow from the objective, in terms of will we consider the objective to follow from the subjective. However, things in the so-called realm of objectivity also appear as an object world of cognition [*ninshiki taishōkai*], if we consider the dependence on the structuration [*kōsei*] of a type of subjective activity, will serves as the act of acts, incorporating the contents of the act of cognition. The contents of will cannot be reduced to the inner [working of] the act of cognition. In this sense of the unique contents of will – in other words the contents of the agency of action [*sayō no sayō no naiyō* 作用の作用の内容] – form the positive contents of will. Time is nothing but an image of the will that the will reflects within the self itself. The category of 'time' arises as a formalization of inner perception when the will engages in introspection of endless depth within the self itself, and the category of 'force' arises when the will itself becomes self-conscious. However, if we see this introspection of endless inner depth from another angle it indicates endless activity, an activity borne independently from within the self itself. The will sees time possessing contents from the perspective of inner perception when it has recuperated an active self. It forms the so-called mental phenomenon. The worlds of mental and substantial phenomena together arise in relation to the *a priori* of the will, while the will arises by seeing the self within the self, so that the will is also an image of the will. But, the contents of what is called the axis of time is somehow to be distinguished. When we consider what appears on the axis of time as merely quantitative, that is, when we consider time itself as simply formal, the physical world emerges; yet when we consider it as qualitative, that is, when we see the contents of time itself, the world of mental phenomena arises. However, though these two together as an image of the will, even when called a world of mental phenomena, incorporate the physical world to the degree to which it may be quantized. Naming it a physical world, moreover, in no way means that it is absolutely not qualitative in nature; all is a local time. If we may, as we have above, consider the world of the physical and the world of mental phenomena as adhering to a single form, and the causality of the natural sciences as forming the basis for modern science, I suppose we may accept something approaching the ideal causality within Plato's philosophy. The ideal [*rinen*] of ideal causality is something developed by the self. That the ideal develops the self itself is to say the ideal sees the self itself. Mental acts must entail the seeing of the self itself.[49] The process by which such an ideal sees the self itself is the movement of time forward. As Plotinus[50] writes, one should consider time as the image of what is eternal. When it comes to the ideal, considered merely on the basis of logic, though it is thought to be unrelated to time, because the one is bound up with [*soku* 即] the many, productive ideals must come into play and must incorporate agency within the self, so that time becomes the form for the acting [*hataraku mono* 働くもの]. Artistic ideals contain activity as if incorporating time; the true ideal must be the creative will, and the ideals which structure the physical world are no exception. The manifest world of the physical realm must be [composed of both] a world of intuition through which the ideal directly sees the ideal itself and a world through which the will sees the will itself. From mechanical causality and final causality[51] to mental causality, all causal categories can be understood as forms by which the will can see the will.[52]

9. Infinite possibilities on the basis of will; the cause of liberty; the latency of time in terms of thought

Here Nishida concludes his essay by clarifying self-awareness and what he means by 'a departure from the consciousness of the active self' when observing the 'object of the self'. Although the terms he uses to describe acting and stillness build on de Biran, Fichte, Lipps and Brentano, the very distinction as refined in terms of acting and not acting implicitly builds upon

Einstein's earliest contribution to not only relativity theory but quantum physics: the observation that the same quanta (called originally the 'quantum of action' (sayōryōshi 作用量子) before being named Planck's constant) used in discussing light in relativity theory at minute scales in quantum physics is predicated on Einstein's clarifications of Planck's work, namely that observable elementary particles behave as both waves in motion and stilled particles. This would in the context of Nishida's other allusions and mentioning of relativity theory suggest the inspiration for associating an act of the will with 'a unity of the acting and not acting' as necessarily 'both moving stillness and stilled motion'. He closes by situating the active self as incorporating limits in terms of will, emotions and ultimately abilities and how we 'consider an infinite range of possibilities' and 'even can negate the actual world'.[53] Nishida concludes by way of what may at first appear to be a somewhat Bergsonian retreat to a moment that defines time as 'a form of functioning creativity', opening him up to continued criticism for aestheticizing the experience of physical and history worlds. However, this essay has primarily argued for a situating of the subject in a way that both affirms free will (and arguably ultimately responsibility for actions) and a responsiveness to the possibilities of worlds we create. It remains one of the most sophisticated and creative engagements with the philosophical implications of early modern physics in any language, and may underscore the importance of how we recognize our worlds in terms of thought and knowledge in time.

Although the world of force forms according to the standpoint of the will, it is not the case that the will itself subsumes itself in a world of force. Will, in terms of the very basis of will itself, takes leave of the world of force and possesses a freedom to run counter to even value systems. Our experience of freedom of will proves it is so. Incorporated in an affirmative judgement is a negative judgement, so that 'some thing is some thing' means that 'it is not some thing that is not that thing'. Thus it seems that in the formation of judgements there must be that which incorporates both of these things from the start. In seeing a given colour as one colour it must be distinguished from other colours. Thus, in forming a pertinent distinction, from the start, regardless of what colours, the formation of these two must occur. As a unity of an infinite number of acts, an act that the will forms accordingly must be a unity of the acting and the not acting, must be both moving stillness and stilled motion. In other words, the unity of an infinite number of actions may even be considered merely that which acts to infinity.[54] However, upon considering the unity of the seen and the not seen, the unity of affirmative and negative actions themselves, then, so as to necessarily consider the seen along with the not seen, the affirmed together with the not affirmed, the will that sees the acting as the object of the self must be both what acts and what does not act. Seen from one angle, knowing too is one type of acting, and it is thought that the will also finds its basis in intellectual agency. At the same time, notably for intellectuals, one can consider the will also as becoming an intellectual object through the course of contemplation. From a purely intellectual standpoint, we can even posit that we disengage from the actual world. We can even say that at the bottom of everything that appears in our mind's eye is a quietly motionless intellectual standpoint. However, when we move a hand, it is not the substance called a hand moving; from where does the consciousness that one moved a hand arise? If seen simply from the standpoint of the intellectual self, the movement of the hand according to one's will and its movement as an external object both must equally be objective phenomena. If both are identical objective phenomena, one of them alone cannot have a special relationship with the self. Insofar as we see a given objective phenomenon as an actualization of the will of the self, the phenomenon must establish a special relationship with the self. The special relationship between this kind of self and the object of the self cannot be simply established through an intellectual standpoint. In terms of viewing the phenomenon of our will, there must be a departure from the consciousness of the active self. That one can conceive of will and emotion as being able to objectify [something] from a standpoint resembling inner perception, in this case, being able to consider inner perception, is due to the self-conscious of the acting self. Given this standpoint, reality and ability are linked. Along with actuality being conferred by way of inner perception, in the background there is always an infinite range of possible things that may be included. Given this standpoint, not only can we detach ourselves from the actual world and consider an infinite range of possibilities, but we can even negate the actual world. Not only can we simply separate ourselves from the actual world, but we can more fully explore the foundations for this standpoint by configuring it within, reflecting it within. Not only can we form limitless possible worlds, but there is even a freedom to form it or not. Underpinning such freedom is the self disengaging itself from the self, the will to lose the will itself, and what appears as possibility *qua* reality is the world of intuition. What is conferred as true must be a world of intuition conferred through a viewpoint of the single suchness of movement and stillness.[55] How we organize our worlds of experience also begins here. What is foundational is always the incorporation of subject–object and form–content oppositions. Here we see, firstly, the infinitely free self inside and the independence of the outside, and according to these themselves we see things in motion. Overcoming time, time is variously cut up in terms of the standpoint of free will that incorporates time within it so as to see various realities. However, if one sees from a viewpoint of the self of single suchness of movement and stillness, whichever self is reflected in the self itself, it can be no more than its image. An ideal sees an ideal itself. Causation in freedom as well as causation in experience both exhibit this same fundamental pattern. As I mentioned above, we see the material world according to an empty time and see the world of mental phenomena according to a time that has contents. According to how rich the contents in time are, one may see various suitable worlds between these two worlds. However, when the ideal truly returns to the ideal itself and the ideal itself becomes self-conscious, so-called actual things become included among possible things, time loses its sense of time itself, and the world of intuition appears. Time is not a form that sees actuality; it is a form of functioning creativity [*hataraku sōzō*]. Herein various worlds of the imagination [*kūsō*] too carry objective significance. In terms of the standpoint of the free self, we can see any variety of worlds all on equal terms. Though we name something the physical world, it is no more than one possible world;[56] we can consider ourselves as freely passing through various worlds. For inner perception to be considered to clearly see all worlds, in a similar sense signifies one's self-consciousness of freedom. Though an intellectual standpoint may be considered to be absolutely removed from time, and while it can be called mere thought, in its capacity as the foundation for objective knowledge it can be called an intuition. Thought, as the basis for all knowledge, must be incorporating the development of limitless acts and in this process potentially incorporates time. The contents of knowledge are imparted from here. Consciousness of what should be constitutes the seeds of time. If we consider the physical world to come into existence by way of the formalization of the contents of time,[57] the world of thought may be said to come into existence by way of the consideration of the potential of time.

Questions for further discussion

1. Why does Nishida repeatedly introduce Fichte's fact-act? Examine each instance of its appearance.
2. How did Nishida include discussion of work in physics by Heinrich Hertz, Rudolf Lotze and Herman Minkowski along with philosophers such as Franz Brentano and Maine de Biran? What key points does Nishida emphasize for each? How are they brought into relation?
3. How might one argue that this essay forms an important step in his development from the subject-centred approach in his first book, *An Inquiry into the Good* (*Zen no kenkyū*, 1911), to the eponymous focus in his essay on 'basho' (site, matrix)?
4. Can an argument be made that while Einstein and Minkowski focused on light to define the limits of causal structurality that had been integral to classical Newtonian physics, Nishida focused on Jamesian pure experience and Zen Buddhist presence at a spatio-temporal nexus to provide his response to special relativity? Is Nishida situating subject and object as such co-variants in light of Einstein and Minkowski?
5. How does Nishida's invocation of Lipps help broaden Nishida's argument? How do they each relate to a Minkowskian world-line approach to time?
6. In developing de Biran's sense of 'convention', Nishida argues that 'Conventions are but the contents of the active self reflected in the world of recognized objects.' How does this differ from how we usually understand conventions and habits?

7. Are disciplinary boundaries within philosophy being challenged or modified by Nishida?
8. Examine a work of art somehow addressing questions in physics. How might Nishida's essay – especially its framing of *poiesis* – form a means of elaborating on the role of will in an artist's choices in constructing the piece(s)?

2.2 Operationalism and the logic of place in Nishida's *Empirical Science*

Introduction

Nishida variously engages **Percy W. Bridgman**'s Operationalist approach to modern physics (long discredited) and Niels Bohr's complementarity, which still remains a formidable contribution among several competing concepts explaining how quantum mechanics may be situated ontologically and epistemologically. Although Bridgman's status today is perhaps easily overlooked in the history of physics and the philosophy of science, like many practising scientists in his day, he was encouraged to write books for general consumption, sometimes converging with other currents of thought (such as logical positivism in Bertrand Russell or emphasis on language use in generating meaning in Wittgenstein). Bridgman's books stirred much debate, as did his steadfast faith in what he called **Operationalism** (see Box 2.2). That Nishida placed such great faith in him is undeniable, but the uses he made of Operationalism suggests a broader reach of issues that spans not only modern physics and philosophical debates over ontologies and epistemologies, but what was called **topological psychology** (see Box 2.4).

> **Box 2.2: Percy Williams Bridgman**
>
> Percy Williams Bridgman (1882–1961) was known not for his work on quantum physics but rather for his work in the field of high pressure physics, for which he was awarded the Nobel Prize in Physics in 1946. However, he penned popular books and occasionally weighed in on debates in quantum physics and relativity as a leading proponent of a controversial approach to any scientific inquiry, which he called **Operationalism**. It demanded that all experiments follow traceable steps in both their conceptual and physical experiment design and processing of results. It may have influenced the formulation of key arguments in the history of quantum physics, even tipping the scales in the pivotal dispute between Niels Bohr and Einstein-Podolsky-Rosen (EPR) over the extent of non-locality ('spooky effects at a distance') and entanglement

of particle angular momentum or spin in quantum physics. That is, Operationalism stood on the side of preserving a sense of human scale in the laboratory setting familiar to classical physics, and as such stood with Einstein in questioning whether quantum physics was complete, the topic of the EPR paper. (Also see Box 3.1.)

For Nishida, it is the disparateness and discontinuous ontologies as understood by way of his own philosophy emerging out of a void-based dialectics and later a worldview (in his later thought) based on a pluralistic intersubjectivity that undermines Western being-based totalizing ontologies, which would have been, at least in an age of classical physics, based on empirical deference to object-constructs. However, with quantum mechanics, both the objects and the spaces and contextualization are problematized, become to some degree unknowable in all cases. To this very extent, quantum mechanics would seem to have the potential to close the postcolonial divide between the hegemony of being-based ontologies in the Western philosophy and nothing-based ontologies in the Kyoto School. Neither *asked* for quantum mechanics, but both find fertile ground for speculation in it. In 'Nishida Kitarō's Philosophy of Absolute Nothingness (*Zettaimu no Tetsugaku*) and Modern Theoretical Physics', Agnieszka Kozyra argues for the influence of experimental physicist Percy W. Bridgman on Nishida's formulations in his 1939 essay 'The Empirical Sciences',[58] which Kozyra renders 'Experimental Science'. In Japanese, *keiken kagaku* 経験科学 certainly refers to 'empirical science(s)', which naturally invokes questions of methodology, and the scientific method is applied in coordination with theoretical suppositions, experiments and results (proofs or refutations of hypotheses or parts of them, etc.). Moreover, in philosophy *keiken-ron* 経験論 and *keiken-shugi* 経験主義 also refer to 'empiricism', reinforcing this use of *keiken*. In any case, that the Japanese word for 'experience' (*keiken* 経験) happens to coincide with the word for the 'empirical' certainly may have helped lead Nishida down the path of a somewhat uncritical embrace of Bridgman's Operationalism, conflating the scientist with an experiential subject.

Indeed, though Bridgman is cited throughout Nishida's essay, reiterating Bridgman's conclusion as a sort of Holy Grail linking everyday experience with quantum mechanics merely repeats, it would seem, the rather embarrassing fallacies evident in Bridgman's surprisingly limited grasp of quantum mechanics. Far from a luminary, Bridgman was in his understanding of quantum mechanics the equivalent of a Newtonian Luddite, approaching a reactionary bent when it comes to presenting classical particle-based physics as a solid, unshakable foundation to which wave physics is appended as an elaborate footnote.[59] In this regard, one may recall Einstein's famous quip describing quantum phenomena as 'spooky action at a distance' as representative of the crux of many quantum mechanical debates related to experimentation and human scales of cognition in relation to the probabilistic world of quantum phenomena. (It should be pointed out that Einstein was wrong, but also that the Standard Model for elementary particles to emerge in the latter half of the twentieth century has recently been complicated by the discovery of the Higgs boson in 2012.)[60] Of course, anyone with a basic knowledge of (the 'new') quantum physics formulated in the mid-1920s knows that its complexity lies stubbornly at its controversial potential dissolution of matter and space as even recognizable, given the complex genesis of matter/waves and entangled nature of elemental particles known according to probabilistic modelling in quantum mechanics. Moreover, basic proofs of concepts of the existence of them today still depend on observations of the *effects* of speculated and observed elemental particles in particle colliders.[61]

Returning to the Kyoto School, Kozyra also presents Nishida as better understanding 'concrete historicist reality' and refutes Tanabe's charges of irrationalism as not standing up because Tanabe was 'stuck in Kantian epistemology' (citing Nishida's own words).[62] This use of Tanabe as a straw man against which Nishida rises by championing a Bridgmanian Operationalism is fraught with problems. First, Bridgman's Operationalism is misrepresented in Kozyra as simply condoned by Einstein and by extension taken rather seriously by others (when it was not). Moreover, Nishida's grasp of quantum mechanics did not reach the level of Tanabe (though Kozyra suggests the opposite to be true). Where Kozyra is helpful is in drawing attention to this essay, 'The Empirical Sciences', and in drawing together the place of intuition in Nishida and the absence of objects and space in quantum mechanics. However, Tanabe has more carefully developed a place for quantum mechanics while obviously building on Nishida's founding concept for the Kyoto School, namely a place for the void in lieu of Being in philosophical investigations.

Translation

from The Empirical Sciences
keiken kagaku 経驗科学, (1939)[63]

1

I wish to write about empirical science from my own point of view. I will consider empirical science in light of my so-called active intuition [*kōiteki chokkan*]. What sort of knowledge does physics study? Though I am not a physicist, my explorations build on the work of Bridgman.[64]

According to Bridgman, the foundational concepts in physics are all operational. It becomes quite evident in the example of Einstein's theory of special relativity. What are called physical concepts too are included in this sense of the physical operations of the active self in the actual world. Still, in the future, they may drift away from operations and perhaps be considered akin to a property of a thing in itself.[65] Time for physicists, independent of physical phenomena, flows in unique appearances, not something resembling a so-called absolute time. It must be a time countable according to physical operations. Time appearing as variables in equations used in physics is a time counted according to clocks. The future is casually considered something synchronous, while in fact it is composed of the complex procedures of physical operations. Synchrony is not a property born of two occurrences; it is relativistic with respect to an observational system. Einstein arrived at this truth by analyzing the physical operations utilized in concrete situations. Though I do not know if he was aware of it or not, but it would follow that he is suggesting that only physical operations can confer significance to a physical concept. In terms of the scope beyond which capabilities of the physical operations cannot go, theory, in a nutshell, will return to the records of the physical operations that are carried out in reality. To this extent, it will never fall into contradiction. Thus, a constituted fundamental physical concept will never require correction. According to the latest experiences,[66] it may be expanded. Yet physical concepts, in considering things like the definition of absolute time or the properties of something removed from physical operations, simply attributing contents [*naiyō*] according to conceptual operations would result in inconsistencies in relation to the experiential. Needless to say, physicists do not stop with the description of actually occurring physical operations, but rather offer postulations concerning the nature of objects and proceed to configure concepts. Then it will be scrutinized through experiments. This process is essential. However, it too must always maintain an operational significance. Whatever remains devoid of an operational significance presents no problem for physics. For instance, even as the size of the earth is constantly changing, as the standard of measurement too is similarly changing, there is no means for knowing it. Such problems are meaningless.

Box 2.3: Standard measure

Standard measure is a translation of a simple term, *shakudo* 尺度, which can usually be translated as 'criterion' or 'measuring stick', but in physics – especially in the context of Operationalism and modern physics in general – forms part of a bridge between the space- and object-centred orientation of physics and its project of inventing laws describing such phenomena, along the lines of Newton's laws that guided classical physics, but which are now subject to difficulties raised by relativity and quantum mechanics. The role of *constants as measuring sticks* for grounding all these emerging modern theories cannot be understated. Without the Planck constant (h), for which Max Planck won the 1918 Nobel Prize for the quantification of energy in h, Einstein may never have developed his relational equation which introduced another constant, the speed of light (the c in $E=mc^2$). Moreover, Planck's constant forms the core of competing theories of wave-based explanations of quantum phenomena, where it appears in its barred form ($\hbar$). (See also see Boxes 3.6 and 3.7.)

Length in physics is measured according to a **standard measure**. When we try to distinguish the length of something, we mark a part of it with a commensurate part of a common measure. Then, moving it linearly until the prior part is brought to the point of the other part. According to the number of times one repeats such a process, the length of an object may be determined. Though such a procedure may seem at a glance simple, in actuality it is quite complicated. It is imperative that the temperature of a standard measure be standardized, and, moreover, if one measures vertical length, one adjusts for gravitational distortions as well. Though complexities arise even in measuring the length of objects at rest, it is all the more so the case in measuring the length of things in motion. In relation to things moving at extremely high velocities, one must put into use different definitions and operations. Einstein's job was to do just that. What Einstein called length and what we call length are not the same. By using the conversion formula from relativity, Einstein combined his approach with ours.[67] Einstein described the world according to coordinate geometry, and what he called length was combined with quantity in terms of formulae in analytical geometry. In measuring the length of something moving, there is a clock that completely unifies all the points in the world; two observers measure the length of something as if measuring something at rest by standing at two ends of it. Such a procedure incorporates that called simultaneity, and simultaneity is relative to movement of the system of observation. Therefore, the length of an object changes according to the velocity of this system. According to how close the system of observation approaches zero velocity, it will approach the length of something at rest. Though the measurement of high-velocity objects resembles the above, when measuring extremely large objects, such as land, we use a surveying instrument. In such cases, the angles of straight lines that join distant points are the angles between rays of light. Here the geometry of rays of light must be assumed to be Euclidean. It is not only when measuring extremely large things that the significance of the operation changes but also when measuring things on the atomic scale. Instead of sense of touch, vision is now used. No matter how small something measured is, no matter how short the wavelength of light that must be used, something resembling X-rays must be used. At such a point we must venture into the theory of optics. An electron's diameter is said to be 10^{-13} cm. In such a case, what is the significance of length? According to how one solves equations concerning electromagnetic fields, one ultimately derives a number.[68] That is because within the concept of length are included theories of electricity expressed according to field formulae. Then, as the range of problems in contemporary physics expands as these and other formulae are proven by way of our experiments, hypotheses are said to be 'correct'. The concept of length too is not independent, but rather maintaining inseparable relations. A change in the significance of an operation here necessitates a change in the significance of what is length.

Because the concepts in physics are said to be operational, the knowledge in physics is all relational. There are no such things as absolute stillness and movement or size. To the degree they are considered to have been determined through operations, they must always be thoroughly relative. However, when quantitative sizes are identical as measured by all people applying the same procedures, it may perhaps be said to be something absolute. Whether it is absolute or not may be determined only through experiments. The absolute is such only experientially and relatively.

 * * *69

How might we explain the ultimate aim of physics? Speaking in terms of a site of operations, there must be a reduction to the elements of what is known at the site of a physical reality [physical situation].[70] In other words, one must present the relations of what is known among phenomena. We always move forward with elements of what is known of things based on previous experiences.[71] Then, we go on to reduce the site of reality to these. However, no matter how our explanations progress with such elements, we at some point find ourselves in a position that no longer permits progress. One such situation occurs when we cannot go forward with our experiment due to reaching an impasse by which the known elements cannot be incorporated into the next stage. One finds this to be the case in gas dynamics. However, here it is merely a matter of the setting being complicated, not the incorporation of new elements. Still, unable to determine the elements of experimental knowledge, we encounter a contradiction in a situation that demands we include, if just minutely, some new element. It presents a crisis of explanation. Such situations, beyond the elements of experimental knowledge, by formally contriving elements that resemble what is local knowledge, perhaps could place all present experience within its mould. However, it would present something merely formal.

Present-day physics is surely confronting a second situation. In quantum phenomena we confront a crisis of explanation. However, such a crisis amounts to no more than some repeated confrontation with something situated in the past. Similar crises of explanation likely confronted Prometheus when he discovered fire or people who witnessed how straw could be glued together by applying friction to amber. We recover the process of explanation by integrating new elements within theories, and do so through the layering of new experiences that amount to knowledge. Our knowledge must all find expression in the language of experience.[72] The range of our experiences is limited. The final stage of explanation always encompasses a periphery. We must seek means to form new bases for explanation by forming new relationships among elements. Quantum physics now stands in just such a situation. With many physicists still exerting great effort to explain the present scope of experience, some are proceeding by devising frameworks out of elements resembling the elements of local knowledge. This now brings us to a third means of going forward, one standing in opposition to the second. Many of the explanations incorporate items from mechanics. This has become a prominent means of thinking. However, it does not go beyond this.

 * * *73

The above roughly presents Bridgman's thought. From the standpoint of operations, moreover he expounds on a wide variety of concepts in physics in great detail. For instance, he is able to consider the grouping of positions in matters concerning space. A thing's position is determined by way of measurements. Therein one must have a concept of length. But concepts of length are differentiated according to operations of measurement. The space by which measurements are made with a standard measure are not identical with them with respect to light waves. Time too is determined according to measurements. Therein the concept of space must enter into consideration. We must bear in mind that there are two species of time: the time of an event taking place at a nearby site, in other words local time, and the time of an event taking place at a distant site.[74] Though the physical operation that measures time has yet to be adequately analyzed, the travel of light between two mirrors forms a standard of measure, and accordingly a unit of time may be determined. However, one must determine the disposition of the light in terms of whether they are still or moving systems, and the possibility that it will move from being a system entertaining, if only slightly, a relative movement of clocks, to another system. At this point all sorts of questions arise. It no longer becomes a simple matter to measure local time without incorporating some measurement of space. But when a measured phenomenon is temporally near, hardly any difference is likely to be measured there. Physicists treat what is called local time as something singular and beyond analysis. Moreover, the foremost concept of causality in physics too incorporates animistic thought. In physics, we do our utmost to leave such thinking behind and set out to hypothesize an independent system based on the repeatability of identical experiments. Then it is hypothesized at length that it will take a certain course. For such a system, consideration is given to the possibilities of external factors and changes being introduced. Of course, in nature such change is inconceivable. There is no such thing as a system that is independent of the physical world. But, as the results of experience such things are conceivably possible. In fact, the isolation of such a system would never be complete, merely partial and approximate. A single event does not have a single effect but rather always countless consequences likely bound up with a countless succession of events. The task for physicists is to try to explain the future as an independent sum of the continuity of events, and to analyze complex causal relations in formative parts.[75] However, whether such analysis is possible always depends upon determinations made in experiments. Formally, such analysis will usually be conceivable as possible; however, operationally speaking, it must exhibit the possibility of fluctuations within the system. If one is to say A is the cause of B, a system within which A does not occur must be experienced.[76] It would be meaningless to inquire about causality if one misses some aspect. In addition, from the operational approach one can expound on its significance in terms various fundamental concepts in physics, including homogeneity, velocity, force and mass, energy, thermodynamics and electricity.

* * *[77]

It is unclear if Bridgman is saying that the context for thought is experience or a physical system or something along these lines; consequently, though I think there are many matters that should be discussed in terms of the relation of concepts with these, one might say that what he said be [situated] in terms of the analysis of physical knowledge itself. Among those who make uncritical scientific assertions, it is rather rare for people to be deficient in their understanding of the role of experience in science. Moreover, building upon the success of science, one extends its authority to supplement one's own arbitrary philosophical decisions.

2

In section two of the essay, one sees Nishida embracing a knowable physical world by way of an Operationalist imagination of means that decline to transcend the physical in the mathematical abstraction of modern physics, but that rather adhere to the visible and objective so that even concepts are treated as concrete assemblages existing in relation to bodily human engagement. It should be pointed out that this approach tacitly serves to stave off charges of abstract remove from history that were levelled against Nishida by Tanabe in 1930, while siding with Bridgman in a way that inhibits a richer comprehension of quantum mechanics.

In Bridgman one may trace a penchant for classical physics and a knowable, controllable world that is reflected in Nishida's adaptation. Bridgman refers to Poincaré's point that 'any aggregation of phenomena, no matter how complicated, is always susceptible of [sic] an infinite number of purely mechanical explanations' to assert a mechanic classical physics view over a mathematical quantum physics view.[78] Thus, he concludes that 'the attitude of the physicist to-day is changing toward mathematical theory' so that 'he takes [physics] far less seriously, recognizes that it contains less of reality and more of a purely suggestive character than he had realized, and lays more emphasis on the demands of simplicity and convenience'.[79] Here, the abstractness of the mathematical orientation inherent to quantum mechanics is somewhat subject to implications of being *unrealistic* insofar as 'it contains less of reality' and *ambiguous* as being 'more of a purely suggestive character', so dependent on statistical probabilities. But most misleading, in relation to Nishida, is the idea that 'our intuitions should be able to tell us what to expect in various experimental situations involving elementary things'.[80] Such an *uncritical* embrace of human intuition goes against the very mathematical orientation of quantum physics, within which geometric and pictorial representations have only a tertiary place.

This turn to psychology seems to return to his first book's approach under the influence of William James, but the focus here remains on a historically-situated approach that reflects developments in contemporary physics. It should be pointed out that Nishida is not a physicist and mixes various discourses and approaches to ontic–epistemological subject–object problems in a dialectic that itself can (and has) been read as being as close to poetry as to philosophy.

What kind of form can an operation take? The question must be considered from the statement that one makes something: *poiesis*. If one says *poiesis*, it may be thought to be simply subjective. However, *poiesis* is not simply the making of a mental image. In the objective world, it is the formation of something external. Things are formed from some elements, and combinations of such elements change. In such outer worlds, to say one makes something necessarily depends upon movement of our bodies, especially movement of our hands.[81] One must adhere to the natural laws of things. Therein must reside something called technique [*gijutsu*].[82] Technique requires our unification with nature. Bereft of such technique, things cannot be made. Though with differing aims, when one makes furniture or the like, it is said that one must make and install instruments for physical experiments. I could not easily identify such a device. In any case, even in making instruments for physical experiments, techniques must be implemented by way of bodily movement. Moreover, physical phenomena are visible by way of such experimental apparatuses.

* * *[83]

Usually, Bridgman writes, basic physical concepts such as Newton's absolute time have been defined in terms of qualities of things independent of physical operations. In philosophy up to now, I believe I have abstractly contemplated the structure of the objective world apart from our formative operations or *poiesis* in terms of an actual historical world. The basic concept is one of providing a definition that accords with a meaning based on conscious operations. And then, when historical reality is, on the contrary, set within such a frame, reality does not follow. Thus what we call *poiesis* too must also entail negation. It is commonplace now that what has until recently formed the frame for physical concepts can no longer accommodate the facts borne of physical experiments. As Bridgman writes on the position in physical reality, our historically formed place, in other words our site of *poiesis*, must have a position [*ichi*] in historical reality. Therein there must be a true concrete reality. The contents of a truly concrete foundational concept in philosophy must be provided through such an operational standpoint. Abstraction and analysis too, no matter what the standpoint, require something of this sort. For persons treating philosophy as a basis for scientific experience, sometime some of them will consider things of a real world isolated from the standpoint of operations. When they mention the world of objectivity [*kyakkan-kai*], they think it simply negates the subjective. We so-called empiricists [*keikenron-sha*] must be all the more rigorous empiricists, and we so-called objectivists be all the more rigorous objectivists.

* * *[84]

That the direct world for us is called bodily is not to say it is an unmediated world preceding the very differentiation of subject and object, but rather that it is a world mediated by way of *poiesis*. To be mediated by way of *poiesis* is to be mediated by technical means. We grasp the world technically. What is called the actual world is something grasped in this manner. Thus, however we proceed we never rid ourselves of this standpoint. Yet, it is not to say that the world is to be treated from a utilitarian standpoint. So-called utilitarianism is nothing more than a standpoint borne of an abstract psychological subject. Moreover, in any sense, what is called conduct [*kōi*] is not considered to come first. What one calls conduct is something arising out of our picturing of the world. The source of our conduct must be a desire. When something other than the self arises out of what is called a self, it is called a 'desire'. If the relationship between things be simply mechanistic or organic, it will not set in motion what we all desire. It occurs out of contradictory self-identity. Thus, according to the techniques used in making something, it may be deemed satisfactory. As psychologists also say, even mental images are necessarily based on impulses. It is why consciousness too is said to be bodily. Yet, to the extent that things are formed technically from it, it is actively intuitive [*kōiteki chokkanteki*]. What becomes the basis for scientific knowledge as experience is also none other than the actively intuited in such a sense. Still, people think of things as if intuitively simply pictured[85] mental images. Thus, to say we grasp the world in terms of techniques is to say we picture a world by actively intuiting based on the elements of the *poiesis*. In this way, the world grasped technologically as forming the self itself cannot be simply said to be either subjective or objective, nor can it be said to be a combination of the two. While the made and the maker, the objective and the subjective are always relative, the world that forms the self itself is always from the made to the maker. It is a world that moves from pattern to pattern. To say a thing can be made in a bodily way, in other words to say that a thing is formed in terms of life [*seimeiteki ni*] is not to say they are the mutual determination of subject and object. The body is not a synthetic compound [*gōseibutsu*]. It must have shifting aspects, both something made and that which makes it.

When it comes to techniques, we usually just consider the subjective, with an individual's skill set in mind. It is considered to conform to that which makes in terms of a world inclusive first of that made and second of that which makes. However, technique necessarily differs from instinct in animals, and there is something that transcends the living body. The specialty of a technique surpasses the living body; on the contrary, it is always something that must be held as a tool. No matter how skilled an architect a beaver may be, to the extent it is based on instinct it cannot be called a technology. For animals, there is no such thing as a skill set. A skill set must entail somebody putting their past to work in the present. While for a technique time is always flowing, it must be established as a self-forming world wherein the past and future simultaneously exist in the present. A technique cannot be formed solely in terms of either a temporal or spatial world. Of course, instinct probably must also be considered in terms of a world wherein the past and future simultaneously exist in the present; however, it may be thought of as merely conforming to a spatial perspective. On the contrary, when one considers it to be conforming to a temporal perspective, it becomes volitional. Neither case has no technique. Technique is present from the perspective of instinct as a result of instinct itself transcending the self itself as self-contradictory, and it is present from the perspective of the will as a result of the will itself transcending the self itself as self-contradictory. When speaking from inside in relation to an effort from inside, force from outside is said to enter into play. The reason that the will is called self-contradictory is that it is established in terms of the utmost limits of such contradictions, as ourselves are desirous depending on what is pictured in the world. Consequently, though will and instinct are forever opposed to one another, what forms a will forms instinct, and the inverse is true as well. An individual who is thought to have mastered a technique necessarily becomes an individual in a contradictory self-identical world wherein the past and future simultaneously exist in the present.[86] The body has its origins in the technical. Thus, though an animal has a bodily existence, it does not possess a body. Consequently, for an animal there is nothing called a world. The world of the present exists where we are technically productive, while as bodily beings we are selves existing in real life. Invoking the technologically produced, there must be form [*katachi*]. It is a form by which the world forms itself, and is a paradigm upon which we depend to function. Therein we have our bodies. The world from which our *poiesis* emerges and towards which it goes sustains self-formatively ourselves themselves.[87]

* * *[88]

When one mentions technological production one usually thinks of something artificial, which as such is thought to not even possess the objectivity of anything... While more than anything we are bodily existences, we have bodies as tools, and herein picture the world, in other words, are conscious. The world that forms the self itself as an absolute contradictory self-unity of the one and the many is a world that is neither simply mechanistic nor expedient,[89] but rather is expressively agential [*hyōgen sayōteki*]. In our bodily formation as historically formative agency [*rekishiteki keisei sayō* 歴史的形成作用], there is what is called expressing the world always as the expressive. Though in the instinctual agency of animals it is probably something exceedingly faint, it cannot be completely refuted on such grounds. As mentioned before, what establishes technology establishes instinct. Therefore, to the very extent we are formative in a technologically and bodily way, the world with perfect clarity reflected gives form to the self itself. Our bodily formation can immediately reveal worldly and historical formations (it is also for this reason that I say the self grasps historical and productive forms in terms of the active and intuitive). Though here one might be tempted to take this thought into the path of aesthetic expression of the world, technologically and productively in terms of the path transcendence of our living bodies, we may say we give rise to so-called mechanical knowledge. As language is also a technology, thought is our freest form of engagement. The historically situated body is always limited by the degree to which it transcends the biological body, and according to these limits the world is expressed in thought. It becomes scientific knowledge. However, while we say scientific knowledge somehow transcends the bodily, it must have some substance [*seishitsu*] by which it emerges and returns from it. It must not lose its operational sense. As absolute contradictory self-identity, historically formative agency is said to have a neutral aspect. Thereby historical formation is always shaken [*dōyōteki* 動摇的], so that I say the historical present is shaken. Therein countless possibilities exist. Therein our consciousness contains endless visions and it is possible for us to say we artificially make things using our bodies as tools. Of course, even so, it is not simply to be said to be up to one's own choice, probably since it is determined historically and socially, as well as a matter of presenting something historically and socially as something objective. However, it is not all set in stone. Usually, only in such circumstances will it be abstractly dubbed technological and productive, while it will without exception emerge from a body as a historical formative agency. It must be something that is incorporated as a moment of contradictory self-identical historically formative agency.

Note how the above positioning suggests a shift from Nishida's earlier work that linked physical engagement at a site with some neo-Kantian aesthetic cognition of his own invention. This can be attributed to a new borrowed (albeit problematic) confidence found in Bridgman that allows Nishida to take a step up from the ambiguities latent in Kantian aesthetics to a lucid reification of a Cartesian clarity in his brand of negative dialectics.

* * *[90]

From that technologically and productively made to what produced it is not to engage, as people say, in a unification of subject and object [*shukaku gōitsu*]. As technologic and productive selves, we have a bodily existence and organic unity as well as use our bodies as tools. We are necessarily technological and bodily, and in addition we have bodies. Though the made [*tsukurareta mono*] is what is separate from us, it still must have bodily form. Without invoking an account encompassing the movement from made to the maker, there is no technological production. When we simply consider such a synthesis as direct, it is no more than the instinctive. One can think of it as an extension of organic bodily operations. It does not follow from the made to the maker. Moreover, even considering as such that which is simply an imaginative agential development of a unification of subject and object would not amount to an account encompassing the movement from made to the maker. One finds no making here. What is made must always be that which arises in relation to the self and the it must always negate the self. It is the same in relation to the self, as it must relate to the other, so that in terms of a common place [*ōyake no basho*][91] there must be a common object [*ōyake no mono*]. As such, the making of something expressively [*hyōgenteki ni*] in relation to the self, on the contrary, must take the form of us being acted upon by that which is expressively in relation to the self. Such a relationship entails a movement from the made to the making. The technical and productive self is what is engaged in the making of such a relationship. Significance emerges concomitant with productivity as the I is said to maintain a body outside and to be bodily in a historical sense. It is not unmediated, and what negates the self itself must always be mediated. The absolute contradictory self-identity must be mediated. Since this always entails mediating that which is other in the self, the body as tool must always mediate a physical world. The I and thou in dialogue do not merely mediate society but must mediate the endless history of the past. However, such mediation may be considered from the self-contradictory self-identity of historical reality [in relation to the movement] from the made to the making. Whatever scientific theory one talks about, all are considered grounded in proof based on experience. At the same time we exist as bodies, our bodies are as tools in our possession too. In other words, calling ourselves historical bodies would seem to only take into consideration everyday commonly occurring events. However, in truth the deep contradictions of human life are found here. Here the endless labours and sorrows of life are found. In making things through techniques we make tools of our bodies and possess a self in terms of what is made. It is such that it always entails a negation of the self as bodily existence. However, it is not simply a matter of it becoming nothing [*mu*]. It is a matter of making in the world of absolute contradictory self-identity, in other words, both a true individual thing [*kobutsu*] and a true self. In always rendering outside inside and inside outside, the self itself is formed expressively [*hyōgenteki ni*]. We can say that our bodily selves as makers take form as the formation of the world itself, which in turn forms the self itself in terms of the absolute contradictory self-identity of [the movement from] the made to the maker. In that the world of absolute contradictory self-identity is considered a historical world, ourselves are born into the historical. Ourselves may be said to be the creative elements of a creative world. However, insofar as the historical world is taken as an absolute contradictory self-identity, it will always be a world thought to be of a simultaneous existence of past and future in the present. In terms of the historical present, we can always be said to exist as individual things in relation to the world. The temporal will always as the maker become spatial, and of course the spatial will be temporal. What is called time is not constituted in a simply linear way as is usually thought, as it is said that time is formed in the present determining the present, it must be spatial in terms of contradictory self-identity. That the maker negates the self itself as the made is a part of the making. Inside is always outside as outside is inside. A relation such as this between self and world may be considered an expressive agency [*hyōgen sayōteki*, lit. expression-agential].[92] In terms of the world of absolute contradictory self-identity, as in the world of Leibniz's monads, along with an individual thing always expressing the world it is an individual thing by virtue of its expression of the world. Conversely, an individual thing may be said to be a means of self-expression of a mutually contradictory self-identity. It is why a monad is also thought to be a perspective of the world from one point of view. From such a standpoint, we may say that we include expression of the world in our operations. The contradictory self-identical world must be understood to express the self itself through innumerable means. Therein one can say innumerable individual things are mutually opposed and engage in mutual determinations through expressive agency. Moreover, on the contrary, the world as a mutual determination of countless individual things can be thought to be moving from something made through contradictory self-identity to a making.

* * *

As Bridgman writes concerning physics, in philosophy too, thus far one thinks of the contents of these fundamental concepts as deriving not from concrete, historical operations but rather from abstract conscious ones. Oppositions between subjective and objective as well as rational and irrational and not to mention their various mutual relationships are all applicable. I do not consider such things indiscriminately, but rather prefer to think of these oppositions and relations from the standpoint of concrete, historical operations.

* * *[93]

As contradictory self-identity [of the movement] from the maker to the made, we see things in actively intuitive ways. To say I am bodily is counterintuitive by everyday thinking. Bodily is what we call the historically operational site of contradictory self-identity, in other words, a technologically operational site. It is not that we consider the historical body as the extension of the instinctual body but rather that the instinctual body is also considered technological. Our bodies must be considered from the starting point of poiesis [*moto-poieshisu yori*]. Consequently, though an animal has a bodily existence, we can say it possesses no body. At the site in which one sees something historically and bodily already, as Bridgman writes, there must be incorporated that known as ingenuity [*kufū*]. Thought in language also is a form of ingenuity. Humans always have a historical bodily existence. In terms of such a historical operational world, the maker and the made are always in opposition, and the world continues to give form to the self itself always as a contradictory self-identity. Usually, it may be thought of as entailing oppositions between subjective and objective or time and space, as the world continues to form the self itself as a contradictory self-identity of space and time or of subject and object. Such formations are not from the one to the many, nor from the many to the one.[94] The I makes use of verbal expression [that moves from] the made to the maker. It may be liable to be interpreted in the causative sense as simply 'from the many'. However, what is 'from the made' is not the same as 'from what is there' [*aru mono kara*]. What has already entered into the past is 'from what is gone' [*naku natta mono kara*]. The made and the making are in opposition in terms of historical space. As contradictory self-identity in this way, historical space always has an inclination towards self-formation. Such space from one side must always be self-expressive.⁹⁵ As bodily historical, we have bodies outside, and in the extreme lose the self itself and simply become expressive. From another side, this self-formation of historical space (as it were, in an opposite direction), insofar as it is always considered self-expressive, should probably even be said, as I sometimes have suggested, to be an expressive universal [*hyōgenteki ippansha*]. All incidents [*dekigoto*] can be thought of as self-determinations of a dialectical universal. Moreover, it is equivalent to a topos or a life-space in topological psychology.⁹⁶ Therefore, in terms of historical space, to the extent that the making and the made in contradictory self-identity may always be considered to exist simultaneously, causal relations may be entertained.⁹⁷ It is a physical space. Moreover, it may be considered a physical world as a world of causal relations in terms of its ultimate truth [*shōgi*].⁹⁸ Material elements in a such a sense must be considered historically operational. In the contradictory self-identical historical space of the one and the many, that all things may be thought to exist simultaneously is a matter of physical position. All events as $B=f(S)$⁹⁹ may be thought to be causal. When historical bodily operations as expressive conscious agency simply become semiotically expressive, contradictory self-identical worlds have lost all consciousness of self-formation, and what emerges is a mathematical world as a world of semiotically expressive operations. Mathematical operations must be historical operations in such a sense. In the physical world, all things may be considered to exist simultaneously, and even taking past and future as negated, still one is left with the opposition of thing to thing in simultaneous existence (still there remains the present in time). In the extreme case of a mathematical world, there is no such thing as time. However, it is not merely a reflection of time being negated, but more likely what could be called the incorporation of time in its abstract transcendence.¹⁰⁰ It is probably due to the contemplation of a mathematical world when considering the transcendence of time in terms of the direction a contradictory self-identical historical world moves in its self-formation and, on the contrary, in terms of the direction that must express the self itself. Thus it would seem a *Mathesis universalis*¹⁰¹ may be contemplated. In considering the past and future as existing simultaneously in the present, from the position of the present being thought to include the past and future, physics, which attempts to provide guidance in all events, is necessarily grounded in mathematics and thus forms a basis for this reasoning.

Physical operations must have a basis in the bodily. Though explanation is said to return the elements of knowledge to a physical location, what are most readily thought of by us as the elements of knowledge are necessarily our bodily movements. 'Bodily movement' does not refer to the bodily or the instinctive of living things, but rather to the poietic [*poieshisu-teki*]. The former on the contrary remains immediately unknown to us. Human bodily movement is poietic; we operate through our historical bodies. It is not just that this is what we know best, but that the thing we call our self [*wareware no jiko*] exists where we make things technologically and bodily [*gijutsuteki shintaiteki ni*] and becomes historical and operational. Moreover, it is there that one locates historical reality. Even the physical operation of measuring the size of something applies a standard measure in a way that must be based on having a sense of its productive significance. Before the advent of physics, within a quantitative world we made things in production-oriented ways. Past and future were considered to exist simultaneously in the present, in a world thought to be a closed world, one bereft of any *poiesis*. However one claims physics to have transcended active intuition, there must be synthesis in *poiesis*, that is, truth requires proof by way of experiments. According to Bridgman, in physical time measurements of space are assumed to take place. Even in saying 'local time' one cannot evade it. In cases when phenomena closely follow a clock, even at some precise time they will be thought incapable of invoking distance. Today too, physicists consider something beyond analysis if based on one sole experiment. However, certainly no one is arguing that distance disappears, just that space-time always must be measured in terms of a contradictory self-identity. Acts of measurement call upon the bodily in history. In the poietic one sees by way of active intuition. Therein it would seem there is a point of departure for a contradictory self-identical synthesis of both the specific [*shuteki*] movement of ourselves as historical processes of formation [*rekishiteki keisei katei*] and expression as its negation.[102] In terms of our bodily movements, functions [*hataraku mono*][103] and seeing [*miru mono*] come to be synthesized in a contradictory self-identity. In a world that is always moving from the made to the making, there must be, on the contrary, a point of synthesis applied somewhere when there is simultaneous existence. Without such points of synthesis, physical experiments would seem impossible. Without assuming intuitional knowledge based on the language of everyday experience,[104] Bridgman also says that one cannot comprehend an equivalence between physical systems and mathematical formulae. Bridgman's writings on formulae mentions that historical formative agency [*rekishiteki keisei sayō*] must be what narrates the self in a self-contradictory way. In so-called classical physics, as the point of departure for physics, our bodily movement would seem to be the model. Even in terms of this point of departure, physical operations are already necessarily active intuitions poietically and in their contradictory self-identity. Still, separated for a long time from operations, they would come to be thought of as the nature of a thing (as would be found in a simple copy theory[105] perspective). Bridgman in the extreme states that the special characteristics of local time that simply defy analysis make us consider in relation to classical physics something resembling absolute time. Physicists have been captivated by Mechanism. Mechanism suggests that there is a binding through our bodily movement. We see things through *poiesis*, having both a bodily existence and our body as tool. To the extent physical knowledge is experimental, it is necessarily rooted here. Our bodies are both biologically instinctive and as historical bodies immediately physical and operational. Physical modelling[106] provides the reason why we play an important role in physics.

Supplement

On Nishida's schematics (a brief overview)[107]

During the productive period that followed Nishida's retirement from teaching he would often append to his collections of essays (which would have already appeared in philosophy journals) new material presenting his thought in schematic formalizations. These appear only in the first three volumes of his seven-volume *Collected Philosophical Writings* (*Tetsugaku-teki ronbun-shū*). Here he would mix physics and topological psychology, ultimately to situate self-awareness in time in the world at sites of emergence and *poiesis* in history. The first such schematic exposition was extensive,[108] elaborating precursors found in even earlier work,[109] while subsequent instalments in volumes 2 and 3 vary in length,[110] but refer readers to the earlier work for more details, suggesting its cumulative nature. Nishida liked to justify his approach in situating the subject by using original relational diagrams integrating a range of ideas, including formulaic expressions from multiple disciplines.

One may compare the symbolic notation used in Kurt Lewin's modelling in *Principles of Topological Psychology* (1936, or earlier publications in German), an obvious influence often cited by Nishida (see Box 2.4). For instance, Lewin demarcates a 'Life space of the individual P' (person) within an environment nested within a sphere described as 'a "dynamically not closed" world' beyond which there is a 'Hull of facts not governed by psychological laws'.[111] Recalling from section 2.1 Nishida's interest in how Maine de Biran situates habitual passive actions and cognition in contrast to active ones, one will see that Nishida was interested less in the codifications of behaviourism or topological psychology, but more in emergence of the ontico-ontological present out of a void as a starting point that offers an alternative to given being. He defines the subject in terms of action. Thus, though he adapts Lewin's formulae '$B = f(S)$' whereby 'B is made an event and S is made a concrete location [*ichi*位置]' and '$B = f(PE)$ expresses all phenomena by rendering [behaviour at a given time B] a function of a person P and the environment E',[112] his aim is not psychological but philosophical, specifically, epistemological in relation to the ontological, and creating a dynamic version of their local co-emergence. Moreover, while Lewin defines environment as '[e]verything in which, toward which, or away from which the person can perform locomotions [as] part of the environment', for Nishida this translates into the language of expressive agency (*hyōgenteki sayō*)[113] that is central to invoking the world and self in contradictory self-identity of things and persons, each the focus in these schematics. Nishida's *basho*-emplaced dialectics redefine the topological boundaries not in terms of struggle with or against a deterministic environment but as co-productive of the person and environment in his unique mutually contradictory dialectics. Nishida attempts to frame acts of situating existence in the world, not to define human psychology per se. Though Lewin also foregrounds the present situation, Nishida's model maintains a physicality of present context by way not of a complex psychology inclusive of blind spots and traumatic blockages or symptoms; rather, it resembles the Minkowski world-lines discussed in the essay 'What Lies Behind Physical Phenomena' (translated in 2.1).

> ## Box 2.4: Topological psychology
>
> Kurt Lewis B is also the function of S, the concrete site of the event. The positioning of a site for an event is also defined by Lewin (and echoed uniquely by Nishida) as $B=f(PE)$ where P is a person and E the psychological environment. Lewin writes:
>
>> Every fact that exists psychobiologically must have a position in this field and only facts that have such position have dynamic effects (are causes of events). The environment is for all of its properties (directions, distances, etc.) to be defined not physically but *psychobiologically*, that is, according to its quasi-physical, quasi-social, and quasi-mental structure. (Lewin, *Principles of Topological Psychology*, 79; emphasis in orig.)
>
> This is certainly the sort of situating of physics, human life and historical context that must have drawn Nishida to embracing and developing Lewin's work, as Bridgman's, as central to justifying and grounding Nishida's own dialectical *poiesis* at sites of person–environment co-emergence.

Nishida explains that 'acting implicates the necessity for a mutually determined mediator for anything independent'. The human mediator is M present at various sites of experience m_n. A indicates the universal, and E the series of e, so that $\frac{E}{A}M$.[114] In such depictions, one can see how Nishida follows Minkowski's world-line elaboration of special relativity theory, which requires a symmetry of emergence of space-time that is more clearly depicted mathematically by coordinate points on the timeline. The segmenting itself surely is made possible by Cantor's limited infinity, the idea that infinity itself is not only literally infinite as embodied in a concept but also as a transfinite set within a set that is infinite in scope. Nishida is not interested in the mathematics of such relationships but merely to sketch satisfying dialectical relationships elaborating on how subject and object are never entirely separate nor merged. The idea of common system-specific structuring of experience of the world by the human subject moving through specific space-times integrates Minkowski as well as Lewin and many other influences on Nishida. As mentioned, Nishida in effect combines a Buddhist void-orientation toward what frames the substratum of existence (instead of being, as in the West), but situates the absolute of space and time in terms of relativity as existing only in isolated sites of emergence in relation to common invariants as determined by the world–self dialectic. This key consideration in understanding Nishida allows us to picture the site (*basho*) in his thought as Buddhist, relativistic and oriented in light of set theory and topological psychology, while also justified in his mind by quantum physics (in his application of Bridgman's Operationalism).

Nishida opens his first instalment of his schematic explanations with the prose: 'To act entails the mutual determination of independent things, and the mutual determination of independent things demands a mediator. Conversely, to act is considered the self-determination of a mediator.'[115] He then introduces the following formulae with the accompanying text (see Box 2.5).

Box 2.5: Subject and object as series in Nishida

$$\frac{m_1, m_2, m_3, m_4, m_5 \ldots}{M}$$

Mediators in relation to things are considered internally and externally. When they are considered externally,

$$\frac{a_1, a_2, a_3, a_4, a_5 \ldots}{A}$$

Nishida explains, 'Considered in this way, a thing takes on the characteristic of something universal, that is its modality. That called the acting disappears.'[116] More specifically, in an earlier preliminary schematic he more clearly defined m as 'a dialectical thing [*mono*]' and M as 'the mediator between particulars [*kobutsu*]', with e indicating 'a particular' and A the universal (*ippan*). There, he also situated M as the mediator between the world of particulars (E) and the universal or ideal (A), so that the mediator determined them as moments in which little m's are produced in a dialectic of the particular and universal.[117]

He then introduces an inanimate schematic (featuring many particular e's in the environment) before introducing an act-oriented one (see Figure 2.1), with the prefatory line, 'In considering the world of the acting, M must be $A \equiv E$', that is, the mediator must embody a correspondence between the ideal (or mental) A and the material E. He later suggests that there are degrees of adherence by the mediator to a balanced 'matter and mind parallelism', 'idealism (or mentalism)' or 'materialism'.[118] Thus, for simplicity's sake, one may say that E indicates given material existence while A indicates mentalism, with M necessarily dialectically engaging them in terms of particular instances of m that themselves manifest mediations of instances of e and a.

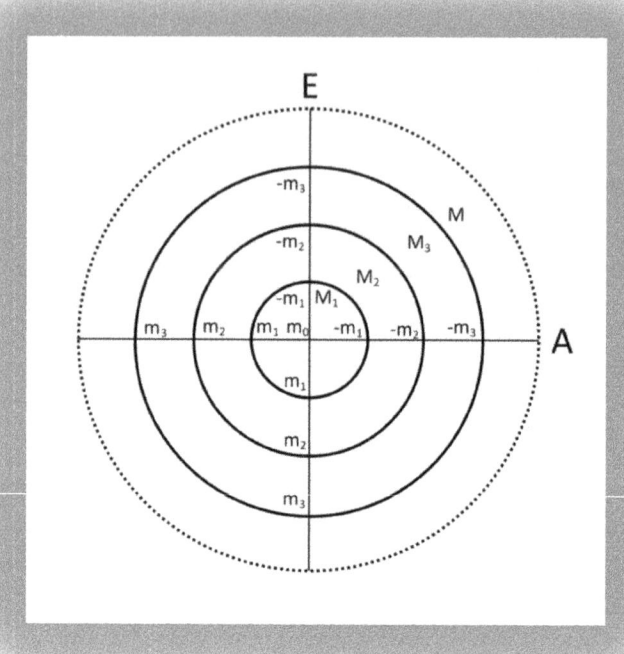

Figure 2.1 Schematic situating the subject.

Note that M (absolute contradictory self-identity) in time (in a Minkowskian sense of timeline or world-line or history) subsumed a change and process as a series M_v ($M_1, M_2, M_3 \ldots M_v$), vaguely along the lines also of a Cantorian transfinite modelling of the local infinite as the world space-time (see also section 4.2). A indicates the ideal, mental or universal (*ippan*),[119] which Nishida associates with tendencies in opposition with the material. For all three versions, Figure 2.1 is central, but in the 1935 and 1937 'Schematic explanations',[120] Nishida includes two diagrams similar to Figure 2.1 but within one the concentric circles are marked by E's and e's instead of M's and m's. Nishida apparently wanted to show in passing that either mediating M or environment E can be placed more or less centrally; however, in keeping with his thesis of promoting a philosophy situating the acting as an invocation of co-emergence with the world, his primary focus is on the schematic with the M's. That he later used only M's and m's also suggests an emphasis on frames of mediation (M), as a turn (or return) to psychology.

Nishida subtitled his 'Schematic Explanations' (*Zushikiteki setsumei* 図式的説明) 'The World of Absolute Contradictory Self-identity' (*Zettai mujunteki jiko-dōitsu no sekai* 絶対矛盾的自己同一の世界). Here, he in effect diagrams his logic of place (*basho*) while integrating his various philosophical, scientific and psychological interests. His variables are apparently abstracted primarily from topological psychology, but his definitions are specific to his dialectical philosophical system. The dynamic modelling provides a placeholder for nothingness and the emergence of becoming in the self and in the world together. Nishida defines the variables and their relations according to various legends and elaborate captions (see Box 2.6).[121]

One last schematic to discuss is the recurring one recreated for Figure 2.2, modelled on the 1939 version, focused on M_x as the axis of a 'world of the historical present that makes things poietically, seeing things actively intuitively'.[122] The 1935 version has one small difference: the M within the outer shell is primed, in keeping with its focus on change over moments. Here, the 'world of absolute contradictory self-identity' considers 'past and future within a simultaneous existence'.[123]

> **Box 2.6: Legend to Nishida's last 'Schematic explanation'**
>
> M = absolute contradictory self-identity (mediator)
> i.e. $A \equiv E$ (congruency between accumulating mental A and material E)
>
> m = individual thing (dialectical thing)
> i.e. $a \equiv e$ (borne of a mediated congruency between instances of mental a and material e)
>
> $E = \sqrt{M}$
> Self-affirmation of the world (individually determined), from the one to the many (E is the circular scope of the determination of material e's)
>
> $A = \sqrt{M}$
> Self-negation of the world (universal determination), from the many to the one (A is the universal, ideal, mental)
>
> Collected from Nishida's 'Schematic explanations' (1935, 1937, 1939).

The following statement reveals Nishida's nested structuring of his dialectical thought and how it shapes the schematics: 'The self-determination of M is absolutely dialectical, and M is dialectical self-identity.'[124] He elaborates that 'Concrete dialectics are not merely a processual dialectics of affirmation and negation, but must be a substantial dialectics of being and nothingness [*umu* 有無]. It must be a true nature dialectics.'[125] Here the emphasis is on moving through time in a dialectics that negates elements in the past and then present while moving into the future. Thus the mediator primed (M') invokes the idea

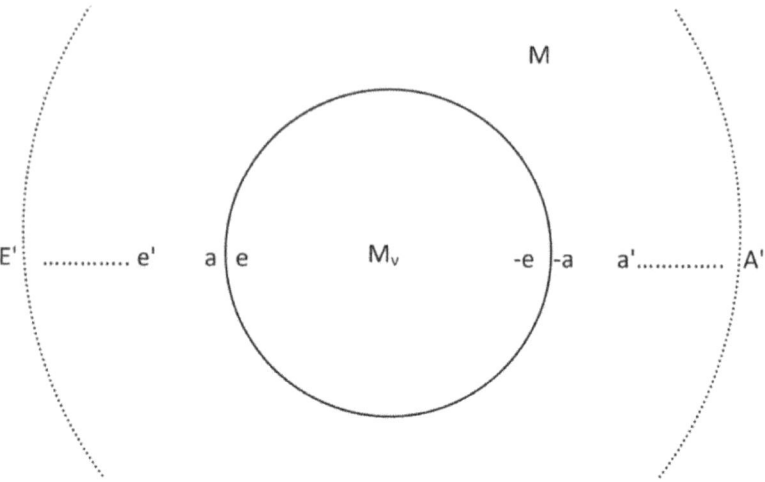

Figure 2.2 A recurring schematic.

of substantial change. Nishida also defines the dialectic of a local site of m in the present as 'e←m→a' so that m serves as the self-determination of M, and m as always serving as an 'expressive agential body' (*hyōgen sayōteki shintai*).[126]

On the one hand, this late-period Nishida Philosophy's neo-Hegelian (and in ways neo-Marxian) recasting of subject–object *dialectics* in terms of sites of creative co-emergence (*poiesis*) out of nothingness embodies clearly Buddhist suppositions as well as relationality derived from modern physics (especially relativity theory) to frame a unique philosophical system. On the other hand, Tanabe's far suppler grasp of contemporary issues in quantum physics exposes (by mere comparison as readers of relevant works) the limits of Nishida Philosophy's dialectical method. Tanabe's work more generally raises awareness of the need to continue to talk about quantum phenomena so that the new relations offered by the language of quantum mechanics may yield new metaphors and naturalized modes of intuition – for instance, the metaphorical language of superpositioning, the uncertainty principle, matrixial positing of particle presences probabilities, wave and field formations, systems and states and relativistic quantum physics (Dirac). This fertile ground for metaphors to elaborate on Nishida Philosophy and his dialectics remained mostly uncharted by Nishida, though Tanabe can be seen to extend some of Nishida's dialectical methods in developing his own approach.

Questions for further discussion

Nishida writes, 'We can say that our bodily selves as makers take form as the formation of the world itself, which in turn forms the self itself in terms of the absolute contradictory self-identity of [the movement from] the made

Nishida Philosophy, Place, Field and Quantum Phenomena 63

to the maker.' Here his reasoning for reversing the usual order of thinking about things as made, in the sense of rendered by the mind, as preceding the object world becomes clearer. He sees the absolute ground for being – both interior subject and exterior world – like Hegel to the extent that mutual negations are invoked; however, instead of moving forward in a socio-historical or encyclopaedic encompassing of the world by the human spirit, a more modest balance of self and world form a static dynamic that was controversial, criticized by Tanabe and others for being overly contemplative. Yet, it can be seen as a means of 'de-Hegelizing' history and the bodily subject without embracing an unoriginal, prepared Marxian system as Tosaka had. It seems to be a reversal of expectations that further entangles the subject, as in *From the Acting to the Seeing*, challenging the norms of classical physics, its epistemologies and its ontologies as in need of what we might call deconstruction (close analysis of its fundamental premises).

Though less pertinent to a focus on modern physics, per se, such issues as how atomic physics disrupts classical concepts of unified space and causality within frames of relation are essential to understanding Nishida's interest in establishing a schematic model for the experientially grounded human subject – not a simple task (see the Supplement). That Bridgman was chosen as an ally is no accident, since he maintains a close proximity with familiar classical (and pragmatic) models. Moreover, given that the long-standing influence of the Copenhagen interpretation – particularly Bohr's focus on complementarity – has itself been attributed to the influence of Bridgman and Operationalism,[127] Nishida elsewhere, as Jacynthe Tremblay has shown, had 'already begun to integrate the concept of complementarity into his philosophy' even before Niels Bohr visited Japan in the summer of 1936. Nishida had found more examples of how nothingness or a void underpins all attempts to ascertain dialectical emergence in space-time.[128]

1. How can Nishida be read as proceeding as a modern scientist in attempting to *objectify* the bodily and worldly mechanisms by which we understand our world? Can any part of his work – some dialectical relation – be reframed or tested as a thought experiment?
2. Nishida writes that 'what we call *poiesis* too must also entail negation. It is commonplace now that what has until recently formed the frame for physical concepts can no longer accommodate the facts borne of physical experiments. As Bridgman writes on the position in physical reality, our historically formed place, in other words our site of *poiesis*, must have a position [*ichi*] in historical reality. Therein there must be a true concrete reality. The contents of a truly concrete foundational concept in philosophy must be provided through such an operational standpoint.'
    a. Here, Nishida is not only applying Bridgman to the physical site of *poiesis* as a human-scaled 'making' but also drawing on the authority of physics as a discipline. Bridgman taught at Harvard and would five years after Nishida cites him here win a Nobel Prize for work in

high-pressure physics. However, the work Nishida cites is Bridgman's more speculative opinions on issues in relativity and quantum physics, opinions solicited by publishers and not very well received or even taken very seriously among physicists working in the new physics. That is not to say he was not influential, since Bohr's approach seems to be influenced by Bridgman or those aligned with him. Thus Nishida uses 'position' in its sense of a position (*ichi*) of points and vectors in modern relativity and quantum physics while placing it within a more historical and less precise and scientific idea of a 'frame' (*waku*), following Bridgman. What does this tell us about Nishida's own criteria for developing a system that is situated in history, inclusive of the history and philosophy of science?

b. How does integrating negation in *poiesis* draw the arts and sciences closer? In our age of digital media, how can Nishida's approach help one situate disparate fields?

3. Nishida writes, 'When they mention the world of objectivity [*kyakkan-kai*], they think it simply negates the subjective. We so-called empiricists [*keikenron-sha*] must be all the more rigorous empiricists, and we so-called objectivists be all the more rigorous objectivists.' One might consider in relation to modern physics how Bridgman and Nishida (and Tosaka, the protagonist of a later chapter, in a different way) preserve a model of external control based on visibility and visualizability that reflects depictions of force reducible to geometric depictions within classical physics. How does their retrenching in this element of force and causality in classical terms form both a defence by which to obviate having to more fully engage the apparent paradoxes of modern physics and their disruption of controlled unified spaces?

4. Try to articulate at more length what Nishida hints at in both Figure 2.2 and his two mentions of the 'single suchness of movement and stillness' in the translation for section 2.2 (see pages 39 and 157 n.55).

5. Nishida integrated many fields in his construction of a system that accounts for a co-productive emergence of self and world. What other fields of knowledge or discourses can you think of that would support this direction? How would you situate them in light of his thought? Would this lead you to amend his schematics in any way?

6. Why might Nishida have felt the need to reduce his philosophical system to such schematics as presented? What is gained or lost?

References

Badiou, Alain, *Second Manifesto for Philosophy*, Cambridge, UK, and Malden, MA: Polity, 2011.

Bridgman, P. W., *The Logic of Modern Physics*, New York: Macmillan, 1927.
Bridgman, P. W., 'Permanent Elements in the Flux of Present-Day Physics', *Science* 71 (1828/1930):19–23, DOI: 10.1126/science.71.1828.19.
Bullough, Edward, '"Psychical Distance" as a Factor in Art and as an Aesthetic Principle', *British Journal of Psychology* 5 (1912): 87–117.
Calvert-Minor, Chris, 'Epistemological Misgivings of Karen Barad's "Posthumanism"', *Human Studies* 37 (2014):123–37, DOI 10.1007/s10746-013-9285-x.
Carroll, Sean, *The Particle at the End of the Universe: How the Hunt for the Higgs Boson Leads Us to the Edge of a New World*, New York: Dutton, 2012.
Cestari, Matteo, 'The Knowing Body: Nishida's Philosophy of Active Intuition (Kōiteki Chokkan)', *Eastern Buddhist*, New Series, 31, no. 2 (1998): 179–208.
Curtis, Robin, 'An Introduction to *Einfühlung* (Empathy)', trans. Richard George Elliott, *Art in Translation*, 6, no. 4 (2014): 353–76, DOI: 10.1080/17561310.2014.11425535.
Fujita, Masakatsu, 'The Scope of Nishida Kitarō's Theory of Place', in Masakatsu Fujita (ed.), *The Philosophy of the Kyoto School*, trans. Robert Chapeskie, and rev. John W. M. Krummel, 13–22 (Chapter 2), Gateway East, Singapore: Springer, 2018.
Kant, Immanuel, *Critique of Pure Reason*, New York: Cambridge University Press, 1998.
Kozyra, Agnieszka, 'Nishida Kitarō's Philosophy of Absolute Nothingness (*Zettaimu no Tetsugaku*) and Modern Theoretical Physics', *Philosophy East and West* 68, no. 2 (2018): 423–46.
Kraus, Oskar, *Oskar Kraus, Franz Brentano. Zur Kenntnis seines Lebens und seiner Lehre, mit Beiträgen von Carl Stumpf und Edmund Husserl*, Munich: Beck, 1919.
Krummel, John W., *Nishida Kitarō's Chiasmatic Chorology: Place of Dialectic, Dialectic of Place*, Bloomington: Indiana University Press, 2015.
Leonard, Lawlor and Valentine Moulard Leonard, 'Henri Bergson', in Edward N. Zalta (ed.), *Stanford Encyclopedia of Philosophy* (Summer 2016 Edition), https://plato.stanford.edu/archives/sum2016/entries/bergson/.
Lewin, Kurt, *A Dynamic Theory of Personality*, trans. Donald K. Adams and Karl E. Zener, New York and London: McGraw-Hill Book Company, 1935.
Lewin, Kurt, *Principles of Topological Psychology*, trans. Fritz Heider and Grace M. Heider, New York and London: McGraw-Hill Book Company, 1936.
Longo, Angela and Daniela Patrizia Taormina, *Plotinus and Epicurus: Matter, Perception, Pleasure*, Cambridge: Cambridge University Press, 2016.
Miyamoto, Wakichi, *Iwanami tetsugaku jiten* (Iwanami's Dictionary of Philosophy), Tokyo: Iwanami Shoten, 1922.
Murdoch, Dugald, *Niels Bohr's Philosophy of Physics*, Cambridge: Cambridge University Press, 1987.
Nishida, Kitarō, *An Inquiry into the Good*, New Haven, CT: Yale University Press, 1990.
Nishida, Kitarō (西田 幾多郎), *Nishida Kitarō zenshū* (Complete Works of Nishida Kitarō), 19 volumes, Tokyo: Iwanami, 1978–9. Abbreviated NKz.
Nishida, Kitarō, *Place and Dialectic: Two Essays by Nishida Kitarō*, trans. John W. M. Krummel and Shigenori Nagatomo, Oxford: Oxford University Press, 2012.

Nishida, Kitarō and William W. Haver, *Ontology of Production: Three Essays*, Durham, NC: Duke University Press, 2012.
Noda, Matao, 'East–West Synthesis in Kitarō Nishida', *Philosophy East and West* 4, no. 4 (1955): 345–59, doi:10.2307/1396743.
Odin, Steve, *Artistic Detachment in Japan and the West: Psychic Distance in Comparative Aesthetics*, Honolulu: University of Hawai'i Press, 2001.
Scribner, F. Scott, 'Nishida's Fichte and the Resistance of Idealism', *International Journal for Field-Being* 1, no. 1, Part 2, Article No. 13 (2001), www.iifb.org/ijfb/FSSScribner-2-13.
Stevens, S. S., 'The Operational Definition of Psychological Concepts', *Psychological Review* 42, no. 6 (November 1935): 517–27.
Tanabe, Hajime (田辺 元), *Tanabe Hajime zenshū* (Complete Works of Tanabe Hajime), 15 volumes, Tokyo: Chikuma Shobō, 1963–4. Abbreviated THz.
Tosaka, Jun (戸坂 潤), *Tosaka Jun zenshū* (Complete Works of Tosaka Jun), 5 volumes, Tokyo: Keisōshobō (勁草書房), 1966–7. Abbreviated TJz.
Tremblay, Jacynthe, 'L'influence du concept de complémentarité dans la philosophie du dernier Nishida', *European Journal of Japanese Philosophy* 3 (2018): 57–77.
Wilhelmsen, Frederick D., *Man's Knowledge of Reality: An Introduction to Thomistic Epistemology*, New York: Prentice-Hall, 1956.

3

Mediation in Tanabe's Dialectical Vision of Competing Fields within Physics

Introduction

Like Nishida, Tanabe was not a scientist; however, he engages scientific writings in great detail, sometimes merely explaining basic concepts in quantum mechanics, but more often – as in the works presented below – relating emerging approaches to growing recognition of unresolved puzzles in relation to classical physics, such as the irreducible duality of matter as both particle and wave (expressed notably as Heisenberg's **matrix mechanics** (see Box 3.10) and **uncertainty principle** (see Box 3.18) and Schrödinger's **wave mechanics** (see Box 3.12)). These formalizations of observed elementary phenomena embrace the dissolution of space-time itself and even matter as commonly conceived. Uncertainty derives from a realization that elementary particles are more wave than particles, more energy than substance. On these questions Tanabe's work excelled. Though sometimes, like Nishida, perhaps overly dependent on one approach over competing ones, he engaged issues in quantum physics in far more technical detail (though neither of them, nor Tosaka, integrate discussions of formulae from physics in their discussions of modern physics).

Tanabe was drawn especially to Paul Dirac, who played a major role in attempts to formulate such a rehabilitation of space by integrating formulae used in both relativistic and quantum mechanical calculations. What becomes most interesting and worthy of a reader's extended investment of time in Tanabe's dense works is his passion for exploring issues in new ways. His original ideas, such as a *logic of species* (or *the specific*) and *absolute dialectics*, while differentiating himself from Nishida's radically self-contradictory dialectics,[1] reflects an age of eugenics and colonial competition of an existential sort both barbaric and somewhat difficult for us to imagine today. However, how he situates quantum physics as formative of a future ontology that was still being elaborated through ongoing scientific investigation remains breathtaking to this day. That these issues are still today unresolved adds poignancy to the fact that he bravely weighed in on some of the thorniest issues in the philosophy of science in his day that are still debated roughly eighty years later. Tanabe in particular raises the issue of how modern physics will (or will not) replace the seemingly comprehensive vision of classical physics with an equally complete new physics. This chapter deepens the comprehension of issues in modern physics (especially quantum physics) and prepares readers for the more general but also detail-rich and exciting issues engaged in the final chapter.

Science, religion, and politics during the height of imperialism and nationalism (1935–45)

One simply cannot avoid questions of politics in the thought of these writers. Tosaka himself engages it openly as a Marxist critic of wartime Japanism, and Nishida increasingly incorporated questions of the historical; however, Tanabe becomes, at the time, an apologist for nationalists in Japan. By bringing these three philosophers into one book, we may notice to differing degrees in all the influence of both Marxist thought – extremely influential among Japanese intellectuals in the 1920s and early 1930s – and physics, which was in the same period both digesting the implications of relativity theory and engaged in intense debates and experimentation in quantum mechanics. One of the key Japan-specific takeaways for world philosophers is a new appreciation of the relation between history and the philosophy of science. The role of science in conjunction with Tanabe's thought on nationalist subjects and moral issues has received scant attention, but it has been noted.[2] In hindsight, one can see that Tanabe seems to have felt compelled to situate contemporary Japan philosophically at the time in the light of Heidegger's sense of the thrownness of being (*Dasein*, lit. being-there) so that his 'logic of species' echoes in the more superficial of senses Watsuji Tetsurō's *On Climate and Culture* and other works that take an unproblematic view of national subjects as givens bound to a land. Such a view may seem extreme today, when America, Germany and Japan are becoming increasingly multicultural societies, but these were givens in the very definition of nations by political scientists for decades. We live in a postcolonial era that still is dealing with racist constructs (racializations) that are the product of the age of imperialism and nationalism that Tanabe in effect bought into in his wartime writings on species. However, as one may explore in the texts translated and situated in the Tanabe section below, a question remains regarding the role of modern science and the philosophical uses to which it was put by Tanabe (and notably ignored by Heidegger in his pre-war writings that influenced Japanese). Thus, an issue in need of exploration (and yet beyond the scope of this introduction) concerns what roles science played in Tanabe's objectifications in his logic of species.

Tanabe Hajime's 'The Philosophical Significance of Quantum Theory' appeared in 1937, during the active period of the Yuiken Marxist research group (and journal) led by Tosaka Jun, who questioned the character of a crisis in modern physics. Tanabe emphasizes that quantum physics has (far more than relativity theory) indeed ushered in a crisis in physics that upsets matter itself by insisting on a simultaneity of wave and particle so that their 'discontinuity according to the parameters of probable waves appears to have been rendered continuous'.[3] Here Tanabe, by merely describing quantum theory, is also casting it in language familiar to the general Kyoto School framing of attention on similar questions concerning a split of temporal flow as a 'continuity of discontinuity', which echoes Buddhist thought regarding all phenomena as transient.

Tanabe even projects onto the contemporary rift[4] among quantum theorists a joy in seeing continuity by way of what he calls an 'intuitive' approach to wave physics, writing, 'Despising the mathematical abstraction of matrix quantum physics as well as its evident discontinuity, physicists as a result found relief in wave physics which could see continuity intuitively.'[5] After summarizing the complementarity theory that renegotiated the divide between wave- and particle-based analyses of quantum phenomena, Tanabe sees the clear result to be that 'wave formulae cannot express causality in the original sense as something determined by relations emergent with the phenomena themselves'.[6] In other words, Tanabe is engaging problems of quantum theory in terms of Kyoto School framing of phenomenology and dialectics as well as how ontologies are understood – and the very radically epistemological shift implied by it. He adds that according to its wave formula-orientation, what is lost is 'a sense of determining temporal processes for phenomena themselves'.[7] Later in the same chapter, he writes that 'quantum theory necessitates a serious limitation on claims of bringing subjectivity to bear in the [pursuit of] knowledge of physics'.[8]

Tanabe applies typical Kyoto School strategies (sometimes coloured by Buddhist terminology) for reorienting a post-Hegelian dialectic capable of explaining quantum theory by arguing that the 'body of the self' is neither objective nor subjective; rather, we are 'the objective rendered subjective and vice versa' because we can no longer depict the differentiation of each.[9] This point itself builds upon both the openness of the void to all possibilities of emergence and quantum mechanics' foundational discovery that the presence of the observer as well as the experimental arrangement will determine if one finds photons sent through a barrier with two openings to behave as waves or particles (this is the gist of the famous double-slit experiment).

3.1 Tanabe's dialectics of classical and modern physics

Introduction

Compared to Nishida, Tanabe's style is dry, yet also elaborately formulated – both syntactically expansive and semantically precise – and rich in details. This piece was first published in 1933, making it clear that Tanabe had been aware of the intense debates following the second wave in quantum mechanics (the 'new quantum mechanics') rather than the earlier one focused more on the relativity theory and understanding the basic structure of waves and particles in the formation of the elemental particles, from 1926 and indeed ongoing. Tanabe refers to the formative developments of Niels Bohr and Werner Heisenberg's Copenhagen Interpretation, which, in brief, demanded recognition of the irresolute nature of quantum observations due to the prevalence of complementarity built into the very quantum phenomena (defined by properties not observable simultaneously, such as momentum and position, and depending on factors such as observation status and experimental apparatus set-up). The Copenhagen Interpretation stood as a barrier to the 'whole' or 'complete' integrity of matter that Einstein sought. It has in the foundations of physics studies within the philosophy of science maintained to this day a firm place as one of several indeed still incomplete but formidable approaches to understanding quantum phenomena (a topic also taken up in the final chapter). The work below exemplifies his appreciation of scientific currents after relativity and quantum physics have altered how philosophy may approach problems of framing phenomena and ontologies. It is a fascinating work that has, like most of the Kyoto School writings on science, been ignored, as attention remains trained on war responsibility and culpability and his relation to religious thought in his late period. Here we find an exceedingly difficult text, one thoroughly engaged in comprehending the implications for space and time, present and locale (the here and now), and matter itself after relativity and quantum mechanics. His primary contribution is an argument that quietly, without mention, suggests a complication of (if not a direct contribution to) a Marxian materialist-dialectical perspective (explored in the next chapter, on Tosaka). Tanabe's very title's inclusion of a '*nature* dialectics' underscores this use of the new physics to undercut the comparatively rough-hewn (how could one be more detailed than the subatomic level) materialism that Tosaka and others offered as 'nature dialectics'. (Related issues of scale are raised in the final chapter.)

Translation

The New Physics and Nature Dialectics (*Shin-butsurigaku to shizen benshōhō* 新物理学と自然弁証法, 1933)[10]

Here Tanabe balances introducing the latest discoveries in physics and developing an original argument that integrates dialogues with continental philosophy in ways specific to the Kyoto School. He asks how the new 'changes in our basis for understanding the natural environment' affect our understanding of nature in the social sciences, especially in Marxian terms. He asks how the implications of the new physics are manifest as a revolution in thought. Tanabe then uses modern physics to demonstrate how changes in time not possible in Galilean transformations of relations in space are integral in special relativity, providing an opportunity to situate more complex spatio-temporal relations between them. Simultaneity itself is no longer possible, but relativity is not about dismissing one position and highlighting another; it offers a model for bringing diversity into spatio-temporal coordination that forms a larger system. Special relativity, seen as such, suggests a joining of limited relata as well as the annihilation of fictions of ubiquitous absolute space or absolute time in isolation (and ironically as infinite).

It goes without saying that relativity theory and the new quantum theory form an axis supporting the new physics. From the time relativity theory has appeared it has constantly been proposing revolutions in physics, which themselves go back over four centuries. Yet today, on the contrary, what may be recognized is that this theory offers the fulfilment of classical physics, so that together they may be said to form an ongoing rendering of these theories themselves as already classical. But indeed a crisis in contemporary physics persists, due to none other than the new quantum physics that has been developed over the last six or seven years. The negation of the law of cause and effect, which may be called the pillar of classical physics, should be seen as the most far-reaching effect of this revolution. As the epitome of scientific precision, these revolutionary theories have shaken the foundations of theoretical physics that is positioned as mediator of how we understand the natural environment; how will we in general now go forward with changes in how we base our understanding of the natural environment? Taken as the foundation of the social sciences, materialist dialectics for Engels even extends to the understanding of nature and advocating a so-called dialectics of nature, which according to such theoretical revolutions in physics are not easy to interpret. This brings us to the problems we are attempting to investigate.

Just as one would expect in interpretations by contemporary physicists,[11] the theory of relativity, in demanding a homogeneity of causal natural description based on the principle of classical physics, may not actually live up to the revolutionary character constantly touted today. When this theory's advocate, Einstein, presents papers, one may likely count no more than twenty people in the world who fully understand them. Though I would not go so far as to dispute it as such, in any case the new thought incorporating this theory is not simply unprecedented; it would seem to demand a complicated change in perspective in terms of its very essence. According to Minkowski, it is the symmetrical beauty of mathematical expressions that draws physicists to applying this theory. Yet even now as it already is becoming classical, in fact, one wonders if physicists may have overlooked the revolution in thought, though it would seem to be integral to this theory. This question will form our first point of inquiry. As is well known, the so-called **special theory of relativity** [see Box 3.2] is the foundation of relativity theory. It is established on two propositions. The first is that there is a law limiting state changes in physical systems: for two coordinate systems in relative motion at uniform velocities, the change in their states are not necessarily correlated. Second, regardless of whether light is emitted from a still or moving object, both possess a fixed and identical speed in relation to still coordinate systems. The second proposition, the hypothesis articulating the so-called unchanging speed of light, it goes without saying, gave theoretical form to the findings of the **Michelson–Morley experiment** [see Box 3.1]. In a similar vein, Lorentz's theory, a supposition concerning the contraction of moving objects (Lorentz's length contraction) explains the results of this experiment according to the hypothesis that there is an inherent local time within the time of a moving coordinate system, in opposition to the absolute time of clocks fixed within an **aether** [see Box 3.1]. Thus, the harmony of bringing together a few theories and experiments negated the role of aether, which had been thought to be the absolute basis for space and time. Thus the first supposition asserts that there is no privileged system that can be singled out among relative, moving coordinate systems; in an entirely relativistic way measurements of time and space are mutually connected according to a fixed transformational form; and for every coordinate system identical natural phenomena may be described by laws of identical forms. Its contents, in terms of substance, demands the relativity of time and space. Its significance is not limited to what we have thus far observed, that space is only concerned with space and time only with time, and does not stop with the assertion that their measurements are relative but that they are what establishes relativity in between space and time. After all, the selection of some starting point for measurements of space is by no means random. Even mutual points of origin in coordinate systems moving relatively at uniform velocities, according to the simple relationship in so-called Galilean transformations, may be joined in what has already long been recognized by classical mechanics. However, in Galilean transformations the transformation of time is inconceivable. Between systems that move relative to one another, time, unrelated to movement, is thus premised upon a sense of it being absolutely fixed. Unrelated to a method by which one somehow will determine synchronicity

Box 3.1: Aether theory and the Michelson–Morley experiment

In casting doubts on the **aether theory** of an infinitesimal yet ubiquitous medium permeating all space, the **Michelson–Morley experiment** is thought to have paved the way for conceptualizing alternative theories of space and matter. It simply found that aether does not exist, shattering assumptions regarding matter and space. It investigated the aether known as luminescent aether, which has been postulated as an explanation for how light can travel. This triggered a complete rethinking of the nature of light, space, and time and moved physics closer to a wave-orientation in attempts to understand matter.

Box 3.2: Einstein's special theory of relativity

Einstein's **special theory of relativity** quantified the speed of light (c) as a constant in a way that Planck's Constant (*h*) had quantified the energy of a photon. Whereas Planck's work formed a foundation for quantum mechanics, the special theory of relativity committed to application that situate larger scales of relationality in terms of light and relative observer positions. This forms a bridge between classical Newtonian physics (that situated mass with causal forces) and what Paul Dirac would draw together into a quantum theory that incorporates Einsteinian relativity (see Box 3.17, on Dirac's quantum mechanics). What is important is that Einstein here provided a means of situating observers in time and space that is necessarily productive of (local) space-time itself, as time and space alone as givens are not possible. Einstein redefines mass (m) itself as bound up with energy that is quantifiable in relation to the speed of light, a leap of understanding that transformed modern physics, liberating it from causality and determinism with a contained whole. Thus $E = mc^2$ became a truism. (See also Box 3.7, Max Planck's energy quanta and Planck constant.)

in some way common to systems that are each in motion,[12] there is nothing but an assumption that simultaneity will be established absolutely. However, relativity theory asserts that simultaneity must be mediated according to transformations depending on their mutual velocities and the intervals between systems that are each in mutual motion and possessing their own characteristics. Thus determinations of simultaneity must still be premised on the condition that a length of a space will be measured. It is due to determinations that match the measurement of the length according to gradations read at a ruler's zero point simultaneously with respect to both ends of the length to be measured. In this way, relative motion of coordinate systems at uniform velocities offers a given moment for participation in the measurements of both time and space. Needless to say, the speed of relative motion at uniform velocities must for both systems have a common value. However, since speed derives from a ratio of space in relation to time, one can not go so far as to claim that it incorporates space and time as components. Consequently, the measurement of space and time is conjectured according to dependence on moving velocities of combinations of both, as self and other, so that space is measured in relation to itself and time, and time is measured in relation to itself and space. Throughout such connections, as velocity seems to be the common value for both, each system must be measured. In addition, given the assumption that the speed of light is unchanging, as well as the need for a unified common value for systems based on the speed of light, in terms of the gap between systems that makes with space and time two relative motions at uniform velocities, they must be measured in terms of relations that would preserve the relative speed and the speed of light of both systems as constants. It follows that dogmatic suppositions like absolute time or absolute motion in Lorentz's theory may now be completely abandoned. Based on empirical facts of completely relative measurement capabilities, on the contrary, the invariant laws of nature are sufficiently guaranteed by means of transformations. The fundamental significance of Einstein's theory of relativity may be considered to exist not in a complete dependence on an empirical theory of relativity but on the contrary in its overcoming.

Above, Tanabe outlines how in modern physics multiple (usually two) coordinate systems may in effect form relations based on Einstein's special relativity theory rather than on Lorentz transformations within a universal aether that would render such systems also placed within absolute space and absolute time. By overcoming an empirical theory of relativity, Tanabe seems to point to the tentativeness in this sense of moving from a universal relationality to a local one. He then continues to refine the implications of this shift and proceeds to assess the extent to which philosophy should modify obsolete assumptions of *a priori* absolute space and absolute time, both of which become incomprehensible assumptions now in light of special relativity theory.

Nevertheless, in order for the theory of relativity to manifest such a significance, new interpretations of space and time that include revolutions in thought of a sort that remains superficial and ill-defined should plainly be extirpated. Although transcendentalists[13] may indeed be called empiricists, I dare say that a person who commands a complete perception of how time is relative to an observer's situated motion probably did not exist until Einstein. The assumption of total measurement that is implied by simultaneity is only meaningful at a single location. It follows that 'the present' is something relative to 'this place', and in relation to systems in relative motion with respect to one another. The recognition of the necessity of changing what we mean by 'the present' would seem to be dependent on his genius. Or one might even consider time as a characteristic of a system and consequently something relativistic, something already asserted in Lorentz's local time. However, Lorentz's local time is a relative time of objects in relation to the absolute time of the aether. As something characteristic of a system, it maintained the absolute time of the static aether that originally related the movements of all systems as its common standard mutual mediator. Thus in terms of this theory, 'the present' is not completely relative with respect to 'the here'. As an image of the absolute 'present' characterized by the absolute space of the aether that mediates all 'localities', it is no more than a reflection of absolute time transcending observation. Measurement of space-time here offers a relativistic facsimile of absolute existence that transcends measurement.[14] However, Einstein's theories completely abandon absolute existence, establishing a standpoint based only on what is empirically measured. In other words, the 'local present' renders a most recent standpoint of cognition in physics. The 'local present' is not simply based on a conceptual determination that establishes it through contemplation. That would only be a practical limitation that would just form in relation to only the body of someone doing something. More concretely, the 'local present' only forms in relation to the actions of observers functioning in nature by means of instruments, which should be identified as bodies and their extensions. For someone simply gazing contemplatively at nature, the 'local present' suggests no significance as a specifically chosen point. Rather, usually what we consider to be a contemplative attitude is as such not simply something passive, but is actually the doing of something [*kōitekinano*], including active movements at the very least in its capacity for making minor bodily adjustments. The 'local present' is something selected in this active manner. This is because actions as points of mediation that exact changes in existence and nature as a commutation point[15] in accord with the human are the 'local present' without mediating it, since some points in time or space cannot enter awareness. It then follows that such a 'local present' is completely characterized by respective observers, so that there is no universal 'local present'. Along with the '*local* present' being relative to the spatial position of the actor, as the 'present' is the present time in which the action occurs, the 'locale' becomes the 'locale' solely on the basis of being relative to the 'present', since the 'locale' first anticipates the active present of the observer that gives meaning to the observer's position. In other words, the 'local' and the 'present' form a unity that cannot be disentangled. Together they structure the moment in which mutual conjectures and determinations are made.

When encountering problems of observers in mutual relative motion, since each observer has a respective 'local present', a plurality of 'presents' simultaneously exists and the 'local' renders the transition of the 'present' from the past to the future into a history. However, only when the 'present' as the 'present' is simply distributed in every point of space, or when the 'local' as the 'local' makes the passage of time into a history, will space and time form an atomistic unity in terms of a 'local time' not dependent upon relativity theory. Classical physics had already begun to have qualms about such issues. Such external combinations of various independent space-times are nothing but arbitrary assertions based on common sense.[16] However, with the quashing of this dogmatism, does not the internal coupling of space-time emerge in some new relationship, in terms of relativity theory's assertion of the relative determinations of their inseparability? As relativity theory would put it, the 'present' is always distinct to the degree it is relative to the 'local', and yet in terms of their simultaneity they are one. Thus if a distinct 'present' is classified as the transition of the 'present' in terms of the history of a 'locale', their coupling must form a 'present' that incorporates another 'present' in mediating a distinct 'locale' in transitioning from one 'present' to another 'present'. In other words, as a single temporal element incorporates the mediation of its determination of spatial elements, in the very process of mediation it will incorporate other temporal elements. Consequently, the alternation of space and time form a temporal (and similarly spatial) self-negating development as well as active unification through splitting and rejoining. It is clearly none other than the dialectical constitution of space-time. Herewith, as a series of 'presents', time may be considered a progression from the past to the future while space may be considered a system of endless possible simultaneously coexisting 'locales'. None of them form independently. Anticipating reciprocity as mutually opposed absolute others, while the one is fractured by negations resulting from the other, on the contrary, they are also negated in turn as the other is contained by the self[17] and restores unity. Thus the coupling of both forms simultaneously.

Like Nishida, Tanabe is drawn to Minkowski's world-line and notes below how it situates not just simultaneity of systems but the temporal element of change as a line from one three-dimensional space into another that incorporates negation in the process of positioning in space-time. Here the dialectical appears. One interesting difference should be mentioned: Nishida Philosophy almost by definition (so integral is it to its formation itself) situates the self-aware human as the local system situated in relativity theory, albeit always somehow entangled with the object world; however, Tanabe refuses this personification or presentation to human scale. Instead, Tanabe remains more faithful to mathematically situated abstract space-time, reflecting quantum investigations in purely probabilistic matrices and the spirit of thought experiments that ultimately has the advantage of recognizing a world closer to a post-human perspective (insofar as the human is not assumed to play the role of the starting point).

As a mathematically expressed dialectical relation of this internalized space-time, **Minkowski's 'world'** [see Box 3.3] displayed epochal significance. After it presented a mathematical expression of Einstein's theory of relativity, for the first time a general theory of relativity could be advanced. Minkowski's 'world' geometry does not simply stop with the mathematical expression of an established theory in physics but rather becomes a 'method' in the development of a new relativistic physics. In terms of such a 'world', space and time themselves are reduced to mere images, while both sustain independence solely through a type of fusion.[18] That is, the fourth dimension of time is no longer simply added to the three-dimensionality of space from outside; rather, spatial elements incorporate the negating function of time into this structure while temporal elements incorporate the negating function of space into this structure. By way of mutual negation, a dynamic inner fusion is achieved, and so a developing unity of the 'world' comes into existence. The 'world points' containing elements of 'worlds' are always elements of movement itself, which drives the self-negating development. It must take the form of a dimension of movement.[19] A 'world' may be thought of geometrically as a curved space, and when the coordinates of time are expressed in imaginary number units, it was likely initially based on the aim of forming symmetry with spatial coordinates in Euclidean spatial terms. In fact, imaginary numbers relate to real numbers by way of dialectical negation, and the negative spatial curvature of hyperbolic space, as if any number of points might form a unity of movements that are relative and mutually warped, and may be considered an indicator for a dialectical structure. Since the 'world' is a dynamic unity of such dialectical development, it permits only the transitory expression of a 'here and now'. Physics accordingly on the whole demands complete adherence to an empirical approach, and thus it preserves this 'world line'[20] all the more tenaciously in relation to the rotation of the coordinate axis of a 'world'. It clearly must be said to indicate a dialectical structure of a 'world'. In the relativity principles of Einstein and Minkowski, space incorporates time as part of its configuration, and time incorporates space in the same way so as to form a 'world' of dynamic unity fused through mutual negation. Through devotion to relativity they on the contrary show the epochal significance clearly present in its dialectical structure as an absolute. In relation to non-dialectical methods of thought such as the law of contradiction[21] in classical physics, this is certainly a completely new transformation of thought. The theory of relativity thus was in this sense, without a doubt, a revolution in physics.

Box 3.3: Minkowski's world

Although Tanabe uses 'world', it may also be referred to as a 'world line', the geometric worlding of a three-dimensional coordinate spatial system along a fourth-dimensional timeline. Although open to interpretation, 'world' in the context of physics and mathematics drawn by Tanabe usually means a coordinate system, a point in relative space that forms an example for how we situate place, time and relationality. Later, his approach may be compared with Nishida's in order to better understand the scope of the Kyoto School in light of their work, including that of Tosaka. Tanabe refers to the pivotal role of Minkowski in Einstein's development – a decade after his theory of special relativity in 1905 – of a general theory of relativity that could include acceleration in the modelling of two moving space-time coordinates. Simply put, Minkowski integrated time as a fourth dimension (line t) added to a three-dimensional (x-y-z) coordinate system, thus forming a 'world' moving with the line points, though not significantly distinguishing itself from ordinary Euclidean space-time. Einstein was then able to demonstrate how space and time are not only co-emergent properties but that they form a continuum that is not rigid and extending ubiquitously but rather subject to forces such as gravity that distort their relation. (Even today, quantum physics continues to recast relativity and gravity itself in the language of elemental particle interaction.)

Tanabe has affirmed the four-dimensional emergence of space-time configurations in Einstein and Minkowski as radically revolutionary in an ontological sense of dialectics that alters not only the present situation (as in Hegel's *Phenomenology of Spirit*) but the very possibilities that classical physics had depended upon (so as to force one to rethink assumptions of both things-in-themselves and *a priori* transcendental categories in Kantian phenomenology that itself was derived in part from Newtonian physics). Unlike Nishida, who uses these developments in modern physics to support his own philosophical system, Tanabe takes great care to define the terms he uses without conflating the human subject and physical systems. In the following passage, he shows precisely how modern physics is revolutionary, which has obvious implications for how we read Tosaka's writings on similar issues. Here Tanabe demonstrates his knowledge of the subtle shifts in situating 'nature dialectics' in light of relativity theory, which depends upon human observers. Moreover, Tanabe sees the pending emergence of a split between physical nature and human that may be addressed in terms of idealist and existentialist thought. Then, Tanabe introduces quantum physics as presenting an even more daunting challenge to classical physics, and goes into details concerning both the foundational scientific discoveries and some of their major implications.

If the essence of relativity theory is understood along the lines of the preceding, its revolutionary significance must be recognized to exist in its introduction of dialectics into theories in physics. At the same time, this dialectic must on the contrary be clearly understood as exceedingly distant from a so-called natural dialectic. That is because a so-called natural dialectic is a dialectic of physical nature that exists apart from the subject, and whereas the subject is something thought merely to render it in terms of its knowledge of physics, relativity theory depends on humans intervening to determine simultaneity as a mediation in terms of measurements of the influence of light in nature, and it would seem to call for the formation of a dialectic in relation to the 'here and now' of this act of measurement. It is not simply a dialectic of substance that is rendered by the subject but rather a dialectic between idealist and existentialist formative steps based on acts that unify the opposites of nature and human. Relativity theory also asserts, along with the establishment of the dialectical structure of the 'world' by clarifying the oppositional unity of space-time in determinations of simultaneity, that this dialectical 'world' is formed only as a rendering of the human 'here and now' of the actively measuring subject as dynamic commutation point. The

> **Box 3.4: Commutational centres**
>
> By '**commutational centres**' Tanabe alludes to commutation in mathematics that may function in relativity theory and quantum mechanics variously as a centralizer, a requirement for central *operators* when dealing with sets of point data as groups and subgroups. Heisenberg later generalized this relationship by way of his enduring uncertainty principle (see Box 3.10). Since relationality between points may no longer be framed in absolute space (or aether), in Tanabe's usage, the centre reflects the set of elements of one 'world' (or system) that commutes (exist in a relation of mutual spatio-temporal definition) with other elements of the 'world' and other 'worlds'.

influence on acts of measurement on nature and its passive rendering become incompatible concepts, so that in the levelling of 'present' and 'locale' in the 'world' forms absolutely relative to subjects placing the 'here and now' as **commutational centres** [see Box 3.4], and bequeaths an inversion of the very essence of a comparatively absolutized nature in the absolute time and absolute space of the aether. Relativity theory does not just seem a basis for a dialectics of nature in such a sense, but rather a dialectic that no longer establishes a materialist perspective based on a theory of reproduction as would be

> **Box 3.5: Newton's law of universal gravitation**
>
> Newton's law of **universal gravitation** states that proportional gravitational relationships between objects are based on their mass and distance from each other. (Contemporary quantum physics generally accepts the yet unproven theory that gravity ultimately is a product of waves of elementary particles dubbed gravitons.)

the case in a so-called dialectics of nature. It necessarily suggests that there must be something idealist-materialist that forms the foundation for the participation of active subjects.

2

Though as a theory of 'worlds' that has dialectical structures along the lines of the above, relativity theory is part of the epochal revolution in physics. Its ken of concern stops at the spatio-temporal structure of the physical world, without approaching the structure of physical existence itself. On the contrary, in light of the inflection of this general relativity on the law of **universal gravitation** [see Box 3.5] and the theory of unified fields, due to its attempt to resituate physical existence into a 'world' space and reduce physics to geometry, it may be considered to completely uphold the pursuit of unified description in classical physics. It is not by accident that **Planck** [see Box. 3.6] considered the modelling of this theory [general relativity] as forming the capstone of classical physics, as one based on a unified and complete physical worldview, from a unified fusion of space-time to the fusion of momentum and energy, to the resituating of the concept of energy in the concept of mass, rejoining measurable mass with inertial mass, even to the point of resituating Riemannian geometry as laws of gravity.[22] However, existence is not something dissolved in a field [*ba*]. The methods of classical physics that demanded the resolution of existence into a field and the resituating of physics in geometry are premised on a non-dialectical direct self-identity of existence; however, it would imply even more the dissolution of physics itself. The revolutionary import of relativity theory will soon be forgotten because the last recognized accomplishment of classical physics has not yet touched on the dialectical structure of physical existence itself. Even in clarifying the relativistic physical world's dialectical structure in the observing subject, it still has not brought a self-negating structure into relation with physical existence itself. Still, this early supposition of revolutionary significance does not go far enough. Rather, in order to explain the results of experiments with radiation, **Planck's** introduction of the hypothesis of **energy quanta** [see Box 3.7], lately developed into a new quantum theory, at last attained a point of developing a dialectical structure for this physical existence. In terms of relativity theory, a dialectic renders the existence of an active

> **Box 3.6: Max Planck (1858–1947)**
>
> A pioneer in modern physics, Max Planck received the Nobel Prize in Physics in 1918 for discovering the quanta of energy that would transform physics.

> **Box 3.7: Max Planck's energy quanta and Planck constant**
>
> **Planck's energy quanta** would be identified as **Planck's constant, h,** which forms a foundational discovery as important to quantum theory as the speed of light is to relativity theory. It allows a quantifiable relation to be drawn between radiation frequency and energy in a photon and led to many related discoveries in atomic modelling and quantum mechanics. It may be expressed in terms of energy as: $E = h\nu$, where ν (nu) is the radiation frequency of a given photon.

oppositional unity visible, not simply a passive rendering of an epistemology grounded in physics. Now, as for the new quantum theory, it clarifies the self-negativity of existence itself in terms of physics. In contrast with relativity theory, it would seem to portend a true crisis in physics.

Planck disagreed with the explanation that had been assumed up until then, that the change in colour that occurs with the rise in temperature of heated radioactive materials is due to continuous fluctuation of evenly distributed energy of a vibrating body with respect to a radioactive object. Instead, radiation must occur in terms of integral multiplications, as each vibrating body has its own inherent units of oscillating energy or quanta, so that the fluctuation of radioactive energy is not continuous but discontinuous and rapid. However, as quanta are proportional to their frequency, their size is distinguished by their distinct vibrating body, and a proportional constant applying to all of them becomes a unifying constant (called h), which indeed hailed a turning point for the century. Since h [see Box 3.7] expressed the product of energy and time, it corresponded to the 'action' of a basic quantum in dynamics. Consequently, quantum theories of energy are necessarily quantum theories of action. Discontinuity of action forms the central ideas in Planck's quantum theory. However, based on research on photoelectric effects, Einstein soon affirmed that light flies through space with the same discontinuity when patterned as energy quanta, and the results of his efforts to collide them with objects produced what appears like photoelectric effects. This was the theory of so-called light quanta or photons. Compton then in terms of photonic theory explained the phenomenon of scattering observed in the collision of the electrons of X-rays, resulting in the definition of the **Compton effect** [see Box 3.8] and overturning the heretofore ascendant wave theory of light, a view that itself had similarly displaced **Newton's old emitter theory of light** [see Box 3.9]. Bohr took a different approach, introducing thought of the active quanta into the structural modelling of atoms. By exploring formula for spectral light according to the determination of non-continuous phases in the energy atoms radiated, he attempted to unlock the secrets of atomic structure by way of quantum theory.[23] At the time, this research took centre place as an outstanding achievement. However, in the Bohr model of the atom, being based on the premise that an electron exhibits continuous

Box 3.8: Arthur Holly Compton (1892–1962)

Arthur Holly Compton played an important role in the development of quantum theory by showing (in 1923) more precisely how light was not only composed of waves but particles, the **Compton effect** (or Compton scattering), for which he won the Nobel Prize in physics in 1927. He later participated in the Manhattan Project to develop the first atomic bombs.

movement of the same sort found in macroscopic movement as determined by classical physics, it selects stationary states of motion discontinuously from among a sustained diversity of orbits of motion for electrons. The model could account for correspondence between these orbits and the quantum nature of radiation energy, and this so-called principle of correspondence would circumvent (the problem of) its original intermediary incompleteness. The result went against the experimental facts, presenting difficulties in explaining the phases of energy in neutral helium, com-

> Box 3.9: Newton's old emitter theory of light
>
> Tanabe uses 'Newton's emission theory' or 'ballistic theory of light', which suggests light is composed of particles, like projectiles from a cannon. It reflects Newton's view in his corpuscular theory of light, seeing light not as atomistic but subject to divisibility, and reflecting their particle by the linear geometry observed in projected light.

pelling the old atomic theory to be necessarily reformulated from the ground up. Thus it is said that for seven years **Heisenberg's matrix mechanics** [see Box 3.10] laid the foundations for the new quantum theory. At that moment, Einstein's theory of relativity shattered the neutrality of Lorentz's theory[24] and in being extended to the relativity of space-time made a comprehensive positivism into a guiding principle. Its significance included forming positions based completely on positivistic facts, so as to return to the spirit of Mach[25] and eliminate the dogma of classical physics persisting in Heisenberg's use of **Bohr's theory** [see Box 3.11]. In other words, in the physics of the microscopic atomic structure, since the movement of electrons and the likes of their orbits, frequencies and so on are not directly observable facts, as something corresponding to them, a theory should be constructed that uses only things like the frequency or energy of radioactive light observed in spectral observations. Thus, one in no way needs a simultaneously completely accurate decision concerning the coordinates and momentum of an electron,

> Box 3.10: Heisenberg's matrix mechanics
>
> Heisenberg's matrix mechanics was a matrix-based vector-depiction of quantum phenomenon that tracks particle positions by eliminating temporal ambiguity. Thus it is also called Heisenberg's picture (or representation) approach. (See also Box 3.4.)

for the estimation of their inaccuracies has already been provided by the **Planck constant, h** [see Box 3.7]. This is the so-called uncertainty principle [see Box 3.18]. Incorporating such an uncertainty principle, along with the momentum and coordinates, a fundamental formula bound together by matrix calculations that did not apply the ordinary algebraic of multiplication in commutative law became essential to Heisenberg's quantum mechanics. Thus, based on this and absolutely independent research in photonics, on the one hand one should not at the same time nullify the existence of the photon and, on

Box 3.11: Bohr's theory of the structure of the atom

Bohr's theory of the structure of the atom was short-lived, yet being visualizable, remains strong in the popular imagination. It depicts atoms as like the solar system with electrons caught in the orbit of the atom nucleus, but fails to explain how they remain or are emitted in terms of physics emerging in this period.

the other hand, take into account the need of the phenomenon of interference in light to maintain a wave nature of light as it has been known. One acknowledges the dual nature of light as both having a particle nature and a wave nature. At the same time, from another angle **Schrödinger's wave mechanics** [see Box 3.12] in the end also corresponded with it, building as it did on the thought of **de Broglie** [see Box 3.13], who endowed matter with both a particle nature and a wave nature. Perhaps, since a so-called matter wave [*busshitsu-ha*] is in relation to wave mechanics nothing but a wave of probability, electrons considered as its wave packet should without trouble be able to infer its correspondence with electrons subject to the uncertainty principle. Yet to the extent that wave mechanics still premises itself on momentum of motion and the coordinates of a position having fixed values at one point in time, it has yet to make a complete break from the thought of classical physics. Thus, in theory, as de Broglie himself recognized, the task of completing quantum physics remained before them.[26] In this way quantum mechanics proceeded about as far as the non-continuous thought of quantum theory, and the new quantum theory would seem capable of completing it.[27]

Box 3.12: Erwin Schrödinger (1887–1961)

Erwin Schrödinger won the Nobel Prize in Physics in 1933 for his foundational innovations in how we understand the wave nature of matter, deriving the Schrödinger equation, the defining formulation of wave mechanics as a *wave function* (before Dirac's work integrated it with Heisenberg's matrix mechanics, which focused on predicting the location of particles). Since Niels Bohr's influential formulations of *complementarity*, which may be seen as a reductive version of Heisenberg's uncertainty principle (see Box 3.18), the Schrödinger equation has been considered to be equivalent with Heisenberg's matrix mechanics. In any case, its far-reaching importance to so many derivative formulae in quantum physics cannot be overstated. One of the most striking aspects of the Schrödinger equation is how it can apply not only to subatomic systems but to the macroscopic manifest world.

Box 3.13: Louis de Broglie (1892–1987)

Louis de Broglie is known for his demonstration that electrons and by extension all matter may be defined by a wave nature, for which he was awarded the Nobel Prize in Physics in 1929. Among the various competing interpretations of quantum phenomena, he is known by way of his 'pilot-wave theory' as the first to allow for the introduction of 'hidden variables' in the formulation of wave-particles (particle-waves). This direction in the foundations of physics was taken up by David Bohm and John Bell, and today spans a drama that is reshaping quantum physics and its history. The influential Schrödinger equation combined his pilot-wave theory to form the basis for a more general wave mechanics (not dependent on hidden variables, which some maintain to be incompatible with quantum physics due to problems posed by Heisenberg's uncertainty principle; see Box 3.18).

$$\lambda = h/p$$

De Broglie defined this key formulation by introducing h (Planck's constant) divided by a value (a defined 'hidden variable') for the momentum p in an equivalency with λ (the wavelength) to redefine how we understand the propagation of waves. Thus classical definition of momentum as mass times velocity ($p=mv$) is replaced by a relation of *wavelength* to define the particle nature of waves or **matter waves**. He also brought *frequency* (f) into relation with the *energy* (E) of a particle divided by Planck's constant:

$$f = E/h$$

(See Tim Maudlin, *Philosophy of physics: quantum theory* (Princeton, New Jersey: Princeton University Press, 2019), 41. (Also see Chapter Five below.)

The most important consequence affiliated with the uncertainty principle in quantum mechanics is its limiting and disavowal of the law of causality in physics, thus endowing this theory with undeniable revolutionary significance. According to the law of causality in physics, a physical object is regulated by physics at a single point in time so that one can decide with certainty determinations in physics in relation to temporal points in proximate succession. Though this was axiomatic, according to the uncertainly principle, since it is unacceptable to specify with precision an object's position and momentum at the same time, the unequivocal significance that determinations in physics demanded of the law of causality is lost, and one can thus only find meaning in relations that exist between probabilities in the ken of the uncertainty principle. The law of causality had been until this point indispensable in enabling accurate descriptions according to differential equations. If one alternatively could still preserve the concept of the law of causality, one would instead spell out interaction in terms of points in time in mutual succession with precise physical determinations, not merely being forced to stop with the definition of successive relations of probabilistic determinations. However, even if this relation of succession is somehow dominant, beyond these determinations in situations in succession themselves stopping at probability, not only must the concept of necessary succession lose most of its significance, but the determinism of the world of physics premised on the law of causality is necessarily robbed of its foundation. As Schrödinger said, unified starting conditions no longer always yield unified results; one cannot avoid the undecidability that comes with stopping at merely comparable statistical results.[28] Nevertheless, since undecidability need not mean being devoid of a law-like nature, while abandoning decidability, physics can establish its position based on statistical laws determined in terms of undecidability.[29] The new physics must realize statistical methods for discovering probabilistic rules for microscopic phenomena. Quantum mechanics too is no different. Yet given that quantum mechanics is endowed with a revolutionary significance, in investigating the source for the unavoidability of the probabilistic point of view, the impossibility of precisely demarcating the means of observation and the phenomena may be termed a point clearly rooted in the difficulty of distinguishing object and subject. Bohr and Heisenberg's commentary is exceedingly clear on this point. Perhaps, in the precise determination of the position of an electron, since light must be shined sufficiently enough on it under the microscope, the light cannot avoid altering the electron's momentum as its energy is transferred to it. The more it tries to avoid it in order to preserve an accurate momentum of the electron, the less one can avoid the incursion of inaccuracies in the determination of position. To borrow a quite pertinent example from Bohr's 'Quantum of Action and the Description of Nature', though when holding a cane one normally is clearly objective in relation to one's sense of touch, upon firmly grasping it and touching things with it, on the contrary the sense of touch becomes capable of locating positions as they are touched by the cane as a physical object, so that just as the cane may be recognized as falling under [the category] of subjectivity, not objectivity, while the radiation of light is objective, as a means of observation it becomes subjective. The duality in these means of observation produce, in other words, a relationship of undecidability. As electrons were first observed in reactions involving such light-shining procedures, their oppositional alterity demanded reforms with respect to the point of view by which means of observation through classical physics were abstracted so that undecidability in the sense of alternating uncertainties was established.[30] Bohr also in this piece cites a collection of papers commemorating the fiftieth anniversary of Planck's doctorate. He writes that comparing the alternativity of aspects originating in this uncertainty with the alternativity of consciousness in psychology, the natural sciences eliminate the epistemological subject on which its unequivocal decidability is based, stopping at limits never actually reached as an independent and complete reality based in objectivity; even the concepts upon which the world of physics is established will never avoid the impositions of the subjective as its conclusions always incorporate uncertainties based on alternatives. As expected, following this direction, one might say that the revolutionary significance of the new quantum theory exists in what was completely overlooked in classical physics: cognition in a mixture of subjective and objective in physical existence in terms of the non-continuous quantum phenomena of microscopic electrons. Physically existing objectivity attempts to refute such an incursion, transforming the subject opposed to the self into an opposition to the nature of the self led within the self itself, beginning to reveal what exists concretely in terms of such a synthesis with a self's opposition, and suggesting the clear triumph of dialectics. The dialectical structure indicated simply in the structure of the world of physics in terms of relativity was established by physical existence itself, which now exists in the world in terms of the new quantum theory. The objectivity of physics also has a dialectical structure by way of its form and contents. As is clarified by relativity theory, the oppositional negation of an epistemological subject does not arrive at an observation through passive objectivity but by active participation, clearly appearing in relation to the mutual uncertainty of the new quantum physics. The complication of subjective and objective in cognition in physics approaches its clearest development with both subject and object forming solely as a unity of opposites. As existence and cognition in physics together have a dialectical structure, the structure is established correlatively and simultaneously through the unification of various oppositional others.

Note how Tanabe closes section 2 with a position remarkably similar, at first glance, to Nishida's own interest in self-contradictory subject-world dialectics, though Nishida gives prominence to self-awareness while Tanabe has focused on how relativity theory and quantum mechanics each offer specific forms of 'dialectics' that demands renegotiations of subject–object relations. Tanabe affirms 'various oppositional others' of multiple discourses impinging on his discussion, including the pressure of Marxian thought quite popular among his students and younger scholars, which had the effect of demanding that historical contextualizations be incorporated in discussions of even the microscopic and cosmological scales of modern physics. Section 3 explore doubts as to whether calling engagements with nature 'dialectical' still makes sense.

3

As elucidated in the first two sections, if we take relativity theory and the new quantum theory as expressing a dialectical structure of nature for physics, the trend in the new physics is dialectical, and one cannot deny the implication that reforms in classical physics are likely needed. In other words, relativity theory starts to require accounting for relative motion at speeds approaching light speed while the new quantum theory raises questions concerning microscopic phenomena such as the motion of electrons, and both of them together fall within the unique domain of physical phenomenon. Yet even if we must approve of the dialectical character of these theories, that they remain but one unique part of physical nature [*butsuriteki shizen*] may give us pause, that we may harbour doubts as to whether this physical nature can be said to be intrinsically dialectical. However, even granting that these apparently unique phenomena in physics being unique only with respect to everyday experience are consequently distant from common sense, in principle on the contrary they are more basic [*kisoteki*] than ordinary everyday phenomena, which cannot encompass the former, while the former can encompass them, since elemental phenomenon are concealed in the totality of effects [*zentaiteki kōka*]. And since nature in physics suggests a whole over which it governs by universal laws, it depends in principle on fundamental laws that for the time being remain distant from everyday experience, the inescapable demand is for the whole to produce a unity. To this extent, one must even be permitted to say that nature in terms of physics in general terms possesses a dialectical structure as a whole. In this way, we probably should endorse the new physicists' continuing progress in the direction of making apparent a dialectics of nature.

In this part, one may note the influence of Heidegger in how Tanabe situates the pursuit of knowledge as an existential act of personal and historical significance positioned at a location. A major question that haunts both Heidegger and Tanabe (as well as Nishida) in the wake of fascism and Japanism that victimized so many non-Japanese (as well as Japanese too, such as Tosaka Jun himself) is the extent to which place is identified with nation and self is identified with national subject. It also has implications for how one situates discrete fields of reference across cultures. Tanabe, as if offering a speculative quantum compromise with Tosaka, outlines a dialectics of nature specific to his philosophical system, yet remaining tacitly in dialogue with Marxists. Note how Tanabe, unlike Tosaka, affirms the work of physicists and up to a point defers to their authority as bases for renegotiating the place of nature in philosophical discourse. Below, Tanabe touches upon history as a topic speaking to the influence of Tosaka and Miki Kiyoshi's materialist critiques of the Kyoto School on him as well as Nishida.

At the same time, here in this so-called dialectics of nature, as has been already variously explained in terms of relativity theory and the new quantum theory, one finds an origination in a reciprocity of subjectivity and objectivity by way of active participation in the epistemological act of observation in physics. Neither of the theories treat light as an object to be understood by physics only objectively; they both place the observing mediator as epistemological subjects in physics with the aim of the observer's body playing a larger role. As a consequence of this duality of the subject and object directly forming the ground for a dialectic, in terms of both relativity theory and the new quantum theory, the dialectical structure appears with the involvement of light. The dual wave and particle natures of light also relate to it in terms of what is considered the dual nature of matter, so that it too may be understood as depending on the dual nature of subjectivity and objectivity with respect to existence in physics. That Bohr, Heisenberg and others as physicists articulately explicated this dual nature would seem to carry tremendously important significance. Accordingly, a dialectics in the new physics indeed has been asserted by physics itself to be based on the alternation of subjective and objective. Seen from another side, the dialectics of the new physics appears as the thinking of a so-called nature dialectics, a dialectical structure of material nature itself independent of and unrelated to subjectivity. This means that since subjectivity does not depict it by way of cognition, it is by way of the active participation of the subject that the object first forms dialectically. It is not objectively material but subjective and objective, conceptual and existential. While the idea of depicting a material dialectical process or structure in terms of cognition cannot be completely confirmed by the new physics, on the contrary, it emphasizes both the active participation of the subjective and the reciprocal formation [*kōgoteki seiritsu*] of the objective. Such is the dialectics of nature required by the new physics. It is neither a dialectics of nature as actual objective existence nor, having said that, a dialectics of the cognition of nature by way of rendered significance. In this way nature and the cognition of nature are separated, with the former being of an independent world existing apart from the latter, and the latter being an image rendering it absolutely passively. From this position, there is no basis on which to establish a dialectics either on the side of existence nor the side of cognition. Only nature and cognition in combination will initiate the formation of a dialectical unity. In neither a separate physical nature nor a rendered cognition are there any unities of opposites. As a moment for precluding the separation of these two, the unification of opposites necessarily incorporates one with the other, along with the expectation that they also variously negate each other in this moment. Indeed, it is through their sublation that it is initially formed. So-called nature and so-called cognition are both thoroughly intertwined, including on the one hand the partial and on the other hand parts of a structure, it supports this formation at a moment of negation. It is on this basis that it may be called dialectical. A dialectical unity suggested along such lines is neither simply of nature nor simply of spirit or cognition. It is an absolute that actualizes the self in terms of the historical existence that establishes the observational acts in the present. Physical nature and mental cognition together form a given moment. Moreover, as we stand with points of view based on scholarly knowledge [*ninshiki*] and active self-consciousness, it becomes impossible to stand directly in a simple nature. Our existence is historical and our cognition engages in active comprehension. A dialectics of an absolute that actualizes the individual in historical existence inevitably appears as a dialectics in our historical, active consciousness. In the instance of this present as the moment of transformation into an active self, what is generalized is nature in physics through fixed relationships mediating the historical existence formed relative to each present self as well as according to mathematical transformations. Though it is a negation of concrete historical existence, it is not an abstract uniformalizing borne of simply negating the present of the self establishing historical existence. The world image of classical physics made Planck's so-called 'liberation from anthropomorphism'[31] its guiding principle. Although it is abstract and uniformalizing, it goes without saying that it is an analytical logical negation governed by law of contradiction, and is not a dialectical negation. In the latter, what is negated must at the same time be preserved as a moment of sublation. According to such dialectical negation, the self in the present is sublated and so-called anthropomorphism is not abstractly but concretely negated, resulting in the establishment of a nature based in physics that is a negation of history. The dialectics of the new physics is none other than such a dialectics of nature. The foundation of dialectics is intrinsically historical. A dialectics of nature is also what renders history constant and uniform in the direction toward nature as this moment of negation and, to this extent, as the foundation for a dialectics of history. In summary, this invarient [*fuhenshiki*] may be called nature. That such a dialectics of nature completely differs in essence from a nature dialectics, I think has already been clearly outlined above. If the absolute that actualizes the self in history as such becomes the foundation for a dialectics of nature, then it is called the nature of reality and said to be dialectical; it must then be something absolutely distinct from a so-called nature dialectics. To regard it as identical with a so-called nature dialectics would be to commit a fallacy akin to asserting that Spinoza's philosophy that explains Spinoza's substance as matter is materialistic.[32]

The final long closing paragraph of this essay further defines issues in relativity theory and quantum mechanics as each themselves in dialectical conflict. What is most suggestive of the depth of Tanabe's understanding of the new developments in physics and their implications for the philosophy of science is the fact that two years after this essay was published, in 1935, Niels Bohr would engage the challenge from Einstein and others, for instance, in the EPR (Albert Einstein, Boris Podolsky and Nathan Rosen) paper entitled 'Can Quantum-Mechanical Description of Physical Reality be Considered Complete?' Thus, one may think of Tanabe's essay as in effect not only engaging issues concerning the limits of Marxism to address the changes in physics, but also as an appreciation of the implications for physicists turned philosophers in the debates over the very nature of physical existence precipitated by relativity theory and quantum mechanics. Tanabe discerned and fleshed out the contours of debates and issues he intuited and yet which had not been yet entered into the Bohr–Einstein debates that continue to shape issues in studies of the foundations of physics. (See Chapter 5 on ongoing debates, experiments and edifying missteps.)

Thinking along the lines of the above, what can be said is that the theory established less by a thoroughgoing relativism in relativity theory than an invariable form of the absolute suggests a guiding principle in relation to philosophical reflection that appreciates the historical invariability of nature in physics.[33] Relativity theory not only simply implies a dialectics borne of a spatio-temporal structure of a world in physics; it also exemplifies a dialectic in relation to the establishment of a physical[34] nature itself. By way of this theory, the Planck school of classical physics had to reconcile itself to radical reforms. Relativity theory does not adhere to the simple completeness of classical thought, and indeed simultaneously entails a negating sublation. As the above would suggest, it possesses a principle significance in relation to the very dialectical formation of nature. With regards to the probabilistic methods of the new quantum theory, for the moment, nature in physics is formed according to an absolutizing and invariant form of relativity theory; it [nature] makes apparent as dialectical character that should not keep concealed either the alternating of subject and object nor contribution of subjectivity, reaching from the general ability of its spatio-temporal 'world' to its materialization in actual existence. Surely probabilities based in statistics express limits due to[35] the collective effects of an infinite number of elements that can in no way each be determined in reality; it is not that they are simply rooted in ignorance with respect to experience but rather that they foreground contingencies by which they are derived agnostically as they are based on principles. Present existence bases itself on contingencies; there is no way to entirely eliminate them. The uncertainty principle expresses them within physics. Determined by relativity theory, the 'world' according to physics, being a 'world' drawn out by 'world lines'[36] of propagated light waves, might well be called simply a 'world' of light. Just as light exists objectively, it also possesses a subjective character as the mediator of the aforementioned observations. Even in terms of physics, it gives shape to the medium in relation to material existence. At the same time, matter on the contrary is both limited to light and gives off light radiantly. Moreover, one would do well to compare this relation as a dialectical relation to the dialectical relation between nation and individuals. Light like a nation is a mediator of the existence of individuals, and only by way of the medium of the nation may individuals exist. At the same time, the nation cannot exist as an abstraction apart from the individuals; just as its contents are always limited by and dependent upon individuals, light depends on matter for its radiance; it is in its dependency on matter that it is limited. Together, as the sublated unity of a mutual opposites, they may be called dialectical. For light in general as expressed both subjectively and at the same time objectively is subject-objective [*shukan kyakkanteki*]; moreover, as in Hegel's sense an objective spirit (its typical example being the nation), while it is a mediator in relation to the existence of the individual it is also made the character of universal existence limited by individuals [*kotai*]. In the new quantum theory, the 'world' of light rendered uniform to the point of physics' demand for a unified schema once dependent on relativity theory once again depends on a dialectics of existence according to physics, clearly displaying a dialectical character according to oppositions with matter. Consequently, though the new quantum theory and relativity theory are variously dialectical, each has a distinct stage and is mutually opposed by way of negation. The latter simply relates to the schemata of possible worlds, while the former relates to the existence of the actual world, and these so-called schemata of possible worlds should not simply be reduced to an abstract form in relation to the existence of the actual world as is contemplated in analytical and logical epistemologies; on the contrary, these are to be unified dialectically, both sustaining mutual opposition, interpenetrating, and beginning to form a concrete existence. Considered a still unresolved problem, the unification of relativity theory and the new quantum theory may be predicted to be in the direction of the resolution. Yet I would not expect these two theories to combine to form a unity in terms of analytical logic. In terms of the dialectical synthesis, each always makes use of the other, and while mutually opposed through negation, moreover requiring sublation in advanced theories. Wondering just what sort of thing their concrete contents are, one all along can only appreciate it as part of the genius of physics. However, there is a dialectical structure to these theories; what alone may be conjectured is that there must after all be a dialectic that concretely dialectically synthesizes the dialectics of relativity theory and the dialectics of the new quantum theory. This pressing demand entails a most concrete dialectics of nature in physics. However, as will already be clear, since dialectics by its very character is a self-negating dynamic unity of subject and object, and nature, bringing formalized invariables to bear on change and laws to bear on contingencies, needs to be formed by a completely objective world of existence, [as it] merely stops at a momentary formation in the historical present, and must be formed anew with each passing present developing in tandem with history. In this sense, nature as an idea will always pose a challenge. A dialectics of nature means a self-negating unity of the subject-objective of nature [*shizen no shukan kyakkanteki*]. As a completely objective existence opposing the subject, a so-called nature dialectics that would assert a dialectical structure – of a nature independent of what would be rendered [*mosha*] by a subject – must be said on the contrary to negate the essence of a dialectic. The dialectical direction of the new physics clearly should be said to prove the failure of a so-called nature dialectics.

Questions for further discussion

This essay appears as one of Tanabe's mediations on the impact of Marxist materialism (and historical context in general) as well as Heideggerian social reorientation (of the context one finds oneself in) that drew Tanabe away from issues solely within the philosophy of science. His first book, published in 1915, introduced contemporary developments in science at a time when quantum theory was still in its infancy – the old quantum theory – which focused on the how to understand the basic structure of the atom. One can see Tanabe reframing materialist concerns over the extent to which modern physics inhibits dialectical engagements, merely takes sides as was evident in the Einstein–Copenhagen rift that in many ways persists to our day. At stake was the question of whether quantum theory was complete in the sense of being capable of replacing classical physics' assumptions of ordinary, manifest human scales of reference; Einstein believed it was not, but made such assumptions based on classical presumption (later disproven by John Bell). If one refused to generalize from quantum phenomena, one denied evidence; if one embraced ad hoc solutions, one limited oneself to a partial account of quantum phenomena. Tanabe seeks 'a dialectic that concretely dialectically synthesizes the dialectics of relativity theory and the dialectics of the new quantum theory', entertaining 'a most concrete dialectics of nature in physics'. He seems to seek a return to a sort of materialism borne within physics, a faith in the dialectical process in physics itself that could contribute to a realization, in quantum terms, of new approaches to ontology and epistemology in philosophy.

1. How do Tanabe's distinctions of the 'local', the 'present' and the 'local present' differ from Nishida's approaches to *poiesis* and *basho*?
2. Reading this essay, does Tanabe adequately situate 'dialectics' in relation to various phenomena outside of modern physics? Place theories from other classes in relation to phenomena specific to modern physics.
3. To what extent might Tanabe be arguing for a dialectical structure in modern physics based on reifications of Hegel? Of Marx?
4. To what extent is Tanabe responding to a Marxist critique (popular still in the Japanese academy at the time) of idealism in general?
5. What are the apparent differences between the dialectical structures Tanabe presents in various scales and at various scopes?
6. How might one argue that Tanabe's emphasis on an 'act of observation (or measurement)' (*kansoku-kōi*) seems to draw not only on Planck's 'quantum of action' but also Nishida's key concept of 'active intuition' (*kōiteki chokkan*), which conjoins thought and action?

7. How does this comparison draw into a common frame quite divergent systematic approaches to questions concerning the mediation of subject and object after modern physics?
8. To what extent can we say that the Kyoto School itself can be defined by issues in modern physics?

3.2 Situating modern physics in the Kyoto School: from Aristotle to Dirac

Introduction

Section 4 of Chapter 6, 'Contemporary Physics and Concepts of Matter in Classical Philosophy', from *Between Philosophy and Science* (*Tetsugaku to kagaku to no aida*), is next presented below. It is representative of Tanabe's most sophisticated treatment of quantum physics attendant to issues raised in the Kyoto School. Whereas some of his works on physics seem to be exploratory or merely introducing to Japanese key concepts in the latest research in modern physics, this work goes much further with economy and confidence. Nishikawa Toshio characterizes the work as 'originally intended to question what the spirit of science is from the standpoint of philosophy, while keeping in mind the national policy of the rousing of this spirit. After stating in the preface that "the recent war harbors a significance that ought to demarcate world history", Tanabe argues that there is a pressing need for a calm "stirring of the scientific spirit" without becoming intoxicated with momentary excitement.'[37] Hideki Mine, after Nishitani Keiji, argues that Tanabe declined to incorporate a 'self-consciousness of absolute nothing' not only because Nishida's 'active intuition' introduces 'misunderstandings', but his 'logic of basho' does not allow for 'differing dimensions or relative domains', in effect inhibiting such 'processual dialectics' from developing.[38] We can understand the general significance of Tanabe's implication that Nishida Philosophy lacks the capacity to include others operative on different fundamental assumptions as a matter familiar to anyone who has considered the framework opened by interdiscursivity and interdisciplinary studies in general. As an introduction to Tanabe, the role of physics has yet to be adequately addressed, since Mine apparently misunderstands the problems Tanabe raises as classical ones rather than specific to modern physics.

Mine mentions the importance of tensor fields in Tanabe's key distinguishing work on the logic of species. In the opening of section 4 of *Between Philosophy and Science*, Tanabe begins with mention of tensors, writing of 'the relation of the relativity of movement' as 'clarifying the inseparability of the existence of matter and the world seen ontologically' insofar as 'general relativity theory ... has succeeded in giving expression to the mechanical conditions of **world tensors** using specifications in relation to the curvature of geometrical space in four-dimensional worlds, and in resolving fields of forces of universal gravitation in **tensor analysis**' (see Box 3.14). Tanabe has certainly pinpointed by focusing on tensors a key meeting ground for philosophy and physics. Tensors can be said to define the very problems raised by both relativity theory and quantum physics in one mathematical framework. Tensor analysis provides one formal means for bringing both perspectives that have been divided by relativity into isolated domains of space-time as well as perspectives that attempt to accrue subatomic models of existence so as to approach more macroscopic scales.

> **Box 3.14: Tensor analysis and world tensors**
>
> In brief, tensors are mathematical means of using matrices (grids of numbers) to quantify complex relations between points in multidimensional (sometimes non-representational) spaces. They are necessary in a world characterized by relativity and quantum issues of determining particle momentum and position.

In relation to Heidegger – who Tanabe disapproved of insofar as he had become a Nazi – does the *patterning* of a site of presence as a state-based context suggest he nevertheless built on a Heideggerian framework of thrownness here? It could even be called in relation to physics a correction of Heidegger's omission of serious treatment of contemporary physics, notably Einstein, in his work *Being and Time* (which seems an omission reflecting Heidegger's anti-Semitism).[39] While Tanabe was interested in liberating thought from the 'closed space of the "modern"',[40] as Mine writes, his reception of Marxism was mixed. Although interested in it and integrating a Hegelian dialectic in the spirit of Marxist historical critique, Tanabe did not embrace key concepts such as '"praxis" in class war' and indeed is closely aligned with Kant and German idealism.[41] On the dialectical in Tanabe, Mine writes, 'For Tanabe, "dialectics" indicates a process that develops through ongoing mutual negation in the mediation of dual moments of oppositions of subjective and objective, consciousness and substance, thought and existence'.[42]

Translation

Section 4 of Chapter 6, 'Contemporary Physics and Concepts of Matter in Classical Philosophy', from *Between Philosophy and Science* (哲学と科学との間, *Tetsugaku to kagaku to no aida*, 1937)[43]

This closing excerpt, from what may be termed Tanabe's magnum opus in terms of his work in the philosophy of science, embraces the new physics in ways that directly engage its implications for the ontological in our world.

Affirming the necessary role of mediation in Aristotle and quantum mechanics, Tanabe turns to Paul Dirac, really the father of quantum mechanics in its mature form, since he united it technically with relativity theory by virtue of his gift for formulaic expression. Aristotle's concept of a substratum serves as a common domain for the macrological relativity and micrological quantum phenomena as Tanabe creatively redefines a common ontology with classical roots.

General relativity theory affects the relation of the relativity of movement in terms of **translational motion** in special relativity theory, and it goes without saying that recent developments reflect progress in expanding its view to include **rotational movement** [see Box 3.15]. It has succeeded in giving expression to the mechanical conditions of **world tensors** with determinations related to the curvature of geometrical space in four-dimensional worlds, and in resolving fields of forces of universal gravitation in **tensor analysis** [see Box 3.14]. It must be recognized that the dissolution of the geometric determinations of the world, as far as the existence of physical objects can be situated in terms of gravity, serve as a synthesis of space-time, a point that holds enormous significance, clarifying the inseparability of the existence of matter and how we see the world ontologically. Instead of comparing space-time in the theory of special relativity with Plato's concept of matter, it might be better to compare it with the inseparability of existence and mediation in terms of Aristotelian substantive matter. Both have in common a continuous point of view that on the surface necessarily results in non-dialectical determinations. Still, the demand for the indivisibility of continuous existence in terms of force fields of matter and polar oppositional discontinuity of affirmation *qua* negation corresponds somewhat to what I have discussed earlier as the indivisibility of a temporally maintained existence of a

---

Box 3.15: Translational motion and rotation

Translational motion depends in Newtonian physics on possibilities of invariance (constants) in relating time translations, through dependency on an assumed homogeneity of time (debunked by relativity), and space translations, through dependency on an assumed homogeneity of space (also debunked), so as to make determinations of energy, momentum, direction and so forth. **Rotation** here is in space and assumes an isotropy of space; it appears also in quantum theory in discussions of angular momentum or spin. (See also Boxes 3.1 and 3.2.)

(See Fritz Rohrlich, *From Paradox to Reality: Our New Concepts of the Physical World* (Cambridge: Cambridge University Press, 1987), 46.)

continuous substratum and, by contrast, a unity of spatial oppositions in a discontinuous 'now'; I wonder if the symmetrical existence of Dirac's yin and yang dual electrons might yet be able to offer an adequate explanation. In physics, of course, the force fields and electromagnetic fields of universal gravitation are seen from the standpoint of so-called macroscopic physics, and one may not conflate them with the microscopic physics of electron and quantum theories. However, from an ontological perspective, so-called macroscopic physics and microscopic physics each exhibit nothing but opposed determinations of a common substratum, so that neither can be absolutely separated or exist apart from the other. Though that may be the case, as the old atomic theory would have it, when dividing macroscopic continuous matter, a limit to its division appears in the form of the indivisible atom, which turned out not to form the elements of existence in microscopic physics. Though such mechanical approaches to atomic theory were already in ancient times established by Democritus, opposition by Plato and Aristotle from an ontological perspective certainly did not permit a simple understanding of idealism's opposition to materialism. They too certainly were not so-called idealists but rather ontologists. Even better, in terms of opposition, in relation to atomic theory thought to offer principles of understanding nature through the movement of atoms in nothingness [*kyomu* 虚無, with connotations of nihility], one might say that dialectical concreteness in pursuit of the indivisibility of existence and space-time is the latent driving force of this ontology. Their so-called indivisible form was not the indivisible elements of matter resembling Democritus's atoms. It has a dialectical form as the negation joined with affirmation in matter. Microscopic atoms are not limited by the divisibility of macroscopic matter; that the opposition of both is uncertain might better be compared in the relation of imaginary and real numbers.[44] One wonders if one might even claim that the ontological significance of the theory of the electron can confirm such an atom based on empirical evidence. The discontinuous atom renders the matter of the world as a continuous substratum a moment of negation; it exhibits a formal existence, in affirmation bound up with negation, that sublates it. Of course physicists are not pursuing experimental research in line with this sort of dialectical ontology. However, when tracing the progress of this research and reflecting upon the ontological significance of their resulting theories, I wonder if it is not to entail just such considerations. The opposition of continuous matter and electrons too, as in the oppositions of gravity fields and electromagnetic fields, or matter and wave, while contradictorily opposed perhaps are inseparable. Might one not also interpret this problem in terms of the relation of a gravitational mass of electrons and an inertial mass? In any case, without a doubt the only appropriate way to interpret the relationship of macroscopic continuous matter and microscopic discontinuous atoms is dialectically. This must include the relationship of waves and particles according to today's new quantum theory. The oppositional inseparability [*tairitsuteki sōsoku*] of material atoms and electromagnetic light waves after all extends to the oppositional bond of waves and particles in terms of matter and radiation, and it may be said to constitute the revolutionary significance of wave mechanics in quantum theory, especially the new quantum theory. While electron theory simply exists as a thing-in-itself [*sokujiteki ni*], in quantum theory indivisibility of continuity and discontinuity can be considered to develop in terms of being-for-others [*taiji-teki ni*]. De Broglie's wave mechanics was deduced by focusing on the simultaneous involvement of waves and particles in physics, and his plan to combine the propagation of waves with the movement of particles into something inseparable formed the grounding for this interpretation. Thus, this theory's derivation of a formula for **matter waves** [see Box 3.13] by applying Lorentz transformations led to the pursuit of a union of world and existence at a level of concreteness not found in general relativity's field theory; it might even be called an attempt to mediate the opposed inseparables of continuous macroscopic existence and atomic microscopic existence as the opposed inseparables of existence and world. Furthermore, one can see it approached a level of thoroughness in Dirac's quantum mechanics. His splendid positron theory [*yō-denshi-ron*] and void hole theory of negative energy and the like may be considered to vividly reflect the dialectical structure mediated by a relativity theory of quantum mechanics.

Above, note the language and issues of Marxian-Hegelian dialectics, Kyoto School Buddhistic ontological roots in void or nothingness and perhaps implications of a refutation or refiguring of Heideggerian being-there (*Dasein*) inspired by the work on Dirac, explored in detail below. Tanabe also writes, 'While electron theory simply exists as a things-in-itself [*sokujiteki ni*], in quantum theory indivisibility of continuity and discontinuity can be considered to develop in terms of being for others [*taiji-teki ni*].' In this sentence one can clearly discern a language that emerges from a Buddhist distinction of self and other within a context of nothingness (or emptiness) as the backdrop for how knowledge of the world is grasped in relation to the self and other in terms of self-power and other-power (*ta-riki*). Two of the terms Tanabe uses – being-for-others (*taiji*) and thing-in-itself (*sokuji*) – may be understood as mapped not only to Hegel but to Buddhism, pivoting on these multiple discursive contexts.

While at the same time wave equations for electrons in quantum mechanics must adapt to the conditions set by relativity theory, the solutions to the equations must take into account the difficulties of representing duality in a way in keeping with the negative numerical values for the energy of motion. In order to circumvent them, Dirac held the mass and charge to be the same as an electron while hypothesizing the marked charge of this positive electron [*sei-naru yōdenshi*], and that the greater the velocity of its movement the smaller its energy, so he considered it unique that at rest it must be supplied energy. However, since in actuality such negative energy has never been directly observed, something along the lines of the following idea must be added. Pauli's exclusion principle attempts to account for all kinds of states of motion according to electrons alone. Accordingly, negative energy states are almost entirely each occupied by single electrons, and then the premise is made that some such similar occupants are unobservable. But, when there is one negative energy state yet unoccupied therein, since an electron that has a negative charge must be added to account for its disappearance, this yet unoccupied negative energy state appears in observations as the existence of a lack of negative energy, in other words, a positive energy [sei-energī]. This is what is interpreted as the positron. As an unoccupied negative energy state, a positron may be compared to a hole. The world possesses a density of electron distribution approaching the infinity of a void [kyo-mugen]. Within a complete vacuum, positive energy states remain entirely unoccupied, as the region is all occupied by negative energy states. The infinite distribution of negative energy electrons plays no part in the electric power field. Thus, the occupation state of positive energy is -e, while each unoccupied state of negative energy is +e in taking their places. Though the exclusion principle forbids the transition of a positively charged electron to a negative energy state in ordinary time, such an electron can appear when it falls into an unoccupied negative energy state. In such a situation the electron [in-denshi] and the positron [yō-denshi] simultaneously annihilate and the energy of both is turned into radiation. On the contrary, electrons and positrons are also created from electromagnetic radiation. The fundamental concept of such a theory clearly maintains complete symmetry of electron and positron in relation to all the laws of physics.[45]

Box 3.16: Orthogonal

Orthogonal means perpendicular in the context of related (rotating) coordinate (x, y) axes. As a concept, it is essential in depicting space-time in relativity theory and matrices and vectors in quantum physics. In the latter, the visualizability of such relations itself is considered by most problematic, being based not on classical relations of force as much as probabilities. X and y vectors maintain a perpendicular relation (are orthogonal) when their inner product is zero, expressed as x⊥y. Thus orthogonality may be understood as a legacy of Newtonian (classical) physics that has been reified in relativity theory and quantum mechanics so as to help conceptualize mathematical vector relations that do not conform to absolute space (space-time is always contingent upon perspective, and requires such a minimalist joining of coordinate systems). Orthogonality explains relations of moving things (particles or waves) and provides a minimal semblance of spatiality and various degrees of temporality based on a rudimentary space-time symmetry between systems as equals, so to speak.

As just stated, Dirac's theory considers the positron as completely symmetrical with the electron and recognizes how through negative opposition it is bound up with matter [*kotai*] existing in the world. Its foundation exists in a dialectical ontology, so that when its actualization is understood in terms of the empirical understanding in physics it is to be deemed of the utmost significance. Regarding his positron as a hole in the world, it is not enough to consider this hole as just a lack of negative energy. On the contrary, like the explanation that has it appear as something positively [*sekkyokuteki ni*] possessing positive [*sei-*] energy, as I mentioned earlier in relation to Aristotle's concept,[46] one may also compare it to what is called lack as existence, which recognizes the positively negative in the actively mechanical. Dirac, for instance, explained the photon's polarization state in general as the superpositioning of two **orthogonal** polarization states [see Box 3.16], and now the intermediary state formed entirely in light of superpositioning does not appear according to results to correspond with the outcomes observed in the original state at all; as one would expect to see the probability observed for the specific results to conform to various equivalent intermediary probabilities in the original state, one may say it clearly presents a dialectical basis for atomic physics.[47] At the same time, it may be thought to offer an extremely suggestive point of view regarding the ontological significance of statistics. Surely even the interpenetration and indivisibility of an ontologically contradictory moment, in light of cognition in physics making mathematical calculations its *modus operandi*, and since relative values of the probabilities of states correspond with various moments, causality in experimental observation is handled as a synthesis of certain probabilities of variously opposed states. Theory attempts to calculate probabilities for experiments to yield specific results. As superpositioning is said to combine the contrary states in terms of such probabilistic observations, it may be understood as none other than a reflection of a dialectical synthesis. Here the dialectical foundation of a probabilistic microscopic point of view is acknowledged. As mentioned earlier, it is the source of the negating opposition to an uninterrupted macroscopic point of view. From this perspective, one may understand the combination of wave and particle not in any classical mechanical sense but rather as depending on mediation by way of probabilities. The microscopic probabilistic theory makes a dialectical ontology its foundation, and it must be realized in light of experimental facts.

Box 3.17: Dirac's quantum mechanics

Dirac's quantum mechanics centres on questions of how to combine relativity theory with problems of locating particles and waves in quantum mechanical terms. His work forms the basis for modern physics today and he was awarded the Nobel Prize in Physics in 1933 along with

Erwin Schrödinger for discovering fresh approaches to atomic theory. Tanabe's interest in Dirac demonstrates his intrepid engagement with some of the thorniest issues in quantum mechanics. The building blocks for his approach concern how to bring coherency that in most mathematical formulations inevitably include forms of symmetry. In Dirac, the Hamiltonian differential equations, Schrödinger's equation and vectors expressed as matrices or indices (Heisenberg's matrix mechanics) figure prominently. Symmetry is expressed through equations that result in commutation or other symmetries between proton and neutron (broken symmetry or approximate symmetry), but it also extends to innovative conceptualizations that led to his postulations of what we today call antimatter. Note that Hamiltonian differential equations (classical in origin but useful in quantum calculations too) function as the starting point for making sense of dynamic relations of energies within systems.

Tanabe's grasp of quantum physics was exemplary. He writes that 'superpositioning is said to combine the contrary states in terms of such probabilistic observations' so as to call for a new 'dialectical foundation of a probabilistic microscopic point of view' that 'makes a dialectical ontology its foundation' and becomes 'the source of the negating opposition to an uninterrupted macroscopic point of view'. The word for 'uninterrupted' (*renzokuteki*) literally means 'continuous', but the sense is in contrast with the broken particle-sized systems of probabilistic quantum mechanics. Tanabe suggests a form of intuition based on the classical manifest world of human scale with the microscopic intuition emerging (still in our time unclear) based on quantum physical observations (see Chapter 5). Note how he brilliantly and seamlessly integrates classical and quantum frames of understanding based on issues grounded in a broader Kyoto School discursive context.

Box 3.18: Heisenberg's uncertainty principle

Heisenberg's uncertainty principle is one of the foundational concepts used to describe (quantify within a probable range of error) what is observed at the subatomic scale in light (particle-wave) behavior. This Principle took a big step in articulating the new quantum logic that broke away from Boolean logic. According to the basic formulae, it can be said to simply state that there is a trade-off between finding the location and the momentum of a particle-wave, depending on how the experiment is set up. The uncertainty in position (Δx) multiplied by the uncertainty in momentum (Δp) must be equal to or greater than the reduced Planck constant (or Dirac constant) $\hbar$, which equals $h/(2\pi)$, over 2.

$$(\Delta x)(\Delta p) \geq \frac{\hbar}{2}$$

If the experiment is set up to find a particle in the moving particle-wave, it will lead to fuzzy results such as the following:

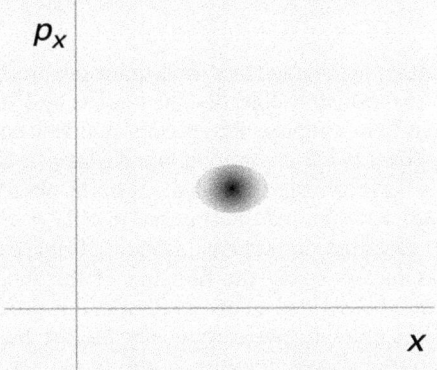

If the experiment is set up to find a wave in the moving particle/wave, it will lead to results over a wide spread of possibilities, such as the following:

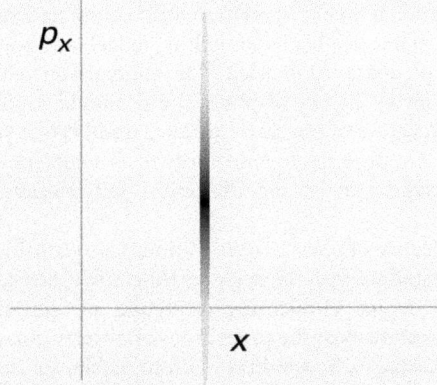

In the wake of relativity theory, which forbids such absolute positioning, and the failure of the Bohr model of the atom to account for the orbits of electrons in an atom according to classical physics, the Uncertainty Principle helps articulate how at subatomic scales electrons and other particles can and cannot be defined in terms of position and linear momentum *at the same time*. It forms a clarification of relations in a quantum mechanical world that must give up the very basis for classical

> Newtonian physics, which strove to quantify all relations of velocity and position by simple knowing variable such as mass and energy in absolute time and absolute space. Subsequent developments in quantum physics often can be traced to the relations Heisenberg brought into focus in this simple principle.

A point that should be further highlighted in **Dirac's quantum mechanics** [see Box 3.17] is that it aims to thoroughly objectify phenomena by dispelling traces of subjectivism seen in Bohr's popular interpretation of **Heisenberg's uncertainty principle** [see Box 3.18]. It is common knowledge that debates around uncertainty, which reflects changes in procedures in the observation of an observed object [*taishō*], have limited the application of laws of cause and effect that are only validated in systems isolated from external disturbances. Interpreted as incorporating the function of the subject as belonging to agency of the mind observing an observation in this situation, I actually must confess to difficulties in overcoming my idealist biases in venturing to offer such interpretations; however, it seems that even the explanations of physicists themselves may to some degree be influenced by such tendencies. Still, in Dirac, thought yielding such faults is uncommon. From a probabilistic standpoint outlined above, while recognizing that disturbances extend external influences to the observed object and why the laws of probability in microscopic phenomena replace the laws of causality, he asserts the necessity of rational theory in relation to the ultimate structure of matter, without taking such uncertainty as something to feel sad about, just the unavoidable structure of observation itself.[48] In comparison with the appearance of subjectivity in the theory of physics, one should say that it presents the self-negating structure of physical existence itself. When you do so, so-called observation is not opposed to the objective as a subjective act, but might be termed the moment of the dialectical self-realization of existence itself.

Speaking from the perspective of physics, observation is also considered a process of the affirmative bound up with the negative [*hitei soku kōtei-teki*] in existence. For physics, with its unrelenting principle of eliminating subjectivity, one expects to restore even the procedures of observation to the incidental structure of existence. It would be unthinkable to do this subjectively. One must express appreciation to my critics for making me see this point. Nonetheless, surely situations of the sort entailing this binding of affirmation and negation in existence are rationalized as the synthesis of probabilistically conflicting moments [*tairitsuteki keiki*]. Thus, philosophically, the uncertainty in the observation of existence in physics should at the same time recognize that being made self-awareness as a gap offering the unrestricted freedom of contemplative subjectivity. Uncertainty in physics should not be said to stem from subjectivity; subjectivity should be said to serve as the mediator of uncertainty. In the contemplation by a subject, the relations of the conflicting moment appearing as the uncertainty of existence are none other than a self-conscious affirmation by way of a negation of negation. Of course, in terms of understanding in physics, since the elimination of the subjective becomes an objective, so in terms of a synthesis of probabilities the negativity and alienation of the conflicting moment is lowered if only minimally, at a glance it looks difficult for such negative unification to achieve self-awareness as free subjects. However, in relation to Dirac's superposition and negation, the probabilistic point of view outlined earlier clearly points to a dialectically unified negative structure. The form inseparable from both negation and affirmation [*hitei soku kōtei naru keisō*] of general material and physical existence forms the contents of contemplation by the subject by means of this negative unity. Furthermore, matter's continuous substratum on the contrary seems to conform to a discontinuous form inseparable from both negation and affirmation, and the entry of the subjectivity of the formal theory as the moment of physical existence itself may be expected to be quite clear when observation actually based on the guidance of theory calls to mind what are necessarily observational results from intentionally prepared experiments. In this sense, theory is the self-conscious side of existence, while thought must be considered the moment of the negation of existence. Existence, being unrelated to cognition, is not the object in depiction, but rather the self-aware existence that renders visible the self in terms of negation joined with affirmation in cognition. Cognition as this self-consciousness, due to the inseparability of self-consciousness and other-consciousness within a dialectical structure, must be the joining of negation and affirmation in existence itself. In this sense, cognition is neither the subjective that configures the objective nor that depicts it, as negation in terms of the objectivity of subjectivity indeed becomes the affirmation of subjectivity; the resulting unification of both in opposition by way of what becomes the moment of sublation of the objective should be said to synthesize cognition as the self-consciousness of existence. Though one may say cognition in physics aims to eliminate subjectivity, it would apply only to the negative side of such cognition. It indeed entails a mediation of subjectivity that affirms the self in terms of negation. If we introduce depiction in this regard, in the sense of what is called depiction as bound up with production in art, depiction must be bound up with production. The uncertainty principle in quantum physics may be made to appear as such a dialectical structure that, coupled with relativity theory, finds importance in fully disclosing the physical world and the negative integrity of its beings. As Dirac himself has been saying all along, the scope for a relativistic development of quantum mechanics has yet to be established.[49] We cannot even begin to predict future developments. However, just in terms of what we now know to be the structure of physical existence, we cannot conceal its dialectical foundations. To this extent, the so-called nature dialectic too must be said to hold a certain currency that should not be dismissed.

In writing that 'the relations of the oppositional moment appearing as the uncertainty of existence are none other than a self-conscious affirmation by way of a negation of negation', Tanabe situates complications in physics due to quantum mechanics in Kyoto School (and perhaps Heideggerian) terms. Note how Tanabe's wording and thought may be contrasted with Nishida's. At such points when the examination of subject–object relations turns to matters of the dialectically situated subject, Nishida will often in his later work situate the subject in terms of self-contradictory subject–object unity, turning inward. By contrast, Tanabe situates the subject more intersubjectively, that is, in terms of how the subject itself is positioned in relation to the exceptional state of solipsistic moments that the contemplative self may find itself falling into. This helps deepen the variety of positions taken by Kyoto School philosophers on Heisenberg's uncertainty principle (not to mention Bohr's further elaboration in the idea of complementarity).

In the past, from the perspective that saw the proper domain of dialectics to be the social sciences not the natural sciences, I claimed that no defendable basis for a nature dialectics existed. However, now I see this assertion must be amended. A nature dialectics demands not only its acknowledgement but the clarification of a dialectical structure of natural existence as well. Theory in the natural sciences also reflects this dialectical ontology and cannot avoid the concept of the affirmative bound up with the negative [*hitei soku kōtei-teki naru gainen*] in determinations of natural existence [*shizen sonzai*]. As in the case of transcendentalist critical philosophy,[50] in actually assuming the existence of absolutely autonomous natural sciences, one demands a subject to synthesize its transcendental foundation. Therefore, the consideration of how to ensure legitimate rights for it entails the abstraction of formalism and conversely the unification of the opposition of science and philosophy. While both maintain various distinct positions, when taken as necessarily becoming inseparable from each other, even the ontological determinations in philosophy become complicated in light of theory in physics, we must recognize how dialectics in philosophy becomes inseparable from mathematical analytical theory in physics. To this extent, a nature dialectic cannot be denied inclusion among theories in physics. That's just not possible. Unlike classical physics, a feature of modern physics as it continues to progress towards the integration of relativity theory and quantum theory must be said to lie in the dialectical structure rendered clear in terms of existence in a world of physics. Thus from the perspective of theory in contemporary physics, one must consider this theoretical imperative of our day through the exacting organization of a nature dialectic. In comprehending quantum physics in all its abstruseness, while wondering how many mistakes I might make along the way, as may even be evident in this scattered essay, I secretly wish to be somehow stimulated by it. Nevertheless, beyond the oppositional conjoining of science and philosophy there is the matter of a needed reconciliation of the mutual opposition of physics and ontology, which certainly may not be seen as identical. Consequently, while physics reflects in its theoretical foundations what becomes dialectical in ontology, at the same time it itself must organize corresponding analytic theories through particular experiments and uses of mathematics. Seen from this side, physics is still an analytical theory and not dialectical. Moreover, to say mathematical existence is dialectical seems to point to a paradox in relation to the infinitely continuous, though having said this is to hastily assume mathematics is dialectical. In other words, [it is] as if one should speak of a result of practical thinking by which mathematics renders non-dialectical what is dialectical, and physics too takes on the mission of rendering dialectical existence non-dialectical. In general, it is said that science has a non-dialectical side by way of its dialectical unification of the dialectical and the non-dialectical. This continuous development occurs as an expansion of the non-dialectical side, a rapid reformation of foundational theory that takes on the appearance of a dialectical aspect. That both are always dialectically unified and inseparable surely corresponds with the dialectical structure in concepts of matter found in classical philosophy.[51] The development of concepts of matter in classical philosophy may be interpreted as foreseen from the perspective of ontology having no aims aligned with the dialectical structure of natural existence as actualized through the empirical understanding of modern physics. Ontologies of the classical age put forth categories for astronomical objects certainly so as to render an ontology of natural existence, likely still significant in ontological form in relation to physics today. On the affirmative side, this nature is even as a living nature remarkable in an Aristotelian way, while on its negative side, in terms of the substratum of matter it corresponds to physical nature. Consequently, there seems nothing particularly odd even though one claims that this ontology takes a form in relation to physics. In this regard, I think to attack interpretations of classical ontologies by means of concepts found in the new physics as anachronistic would be to draw a conclusion without understanding the inseparability [sōsoku] of ontology and empirical cognition. On the contrary, to contemplate and philosophically mediate this inseparability, whether for classical ontological interpretations or critical contemporary physics, would in either case seem to carry significance. From critical philosophy's standard view of the relationship of philosophy with the natural sciences, one simply sees from the angle of the latter's discovery of transcendental foundations for conceptual configurations. From this ontological viewpoint one sees a unity of the opposition of phenomenal reality and an essentialist structure of existence that is considered to be of the highest importance today in order to avoid the dogmatism of the old natural sciences and at the same time overcome the formalism of critical philosophy. It may be said that philosophy and the natural sciences give rise to a reciprocal relationship of form [keisō] and matter [shitsuryō] concretely understood and certainly in itself ontologically seen. A nature dialectic is nothing other than the dialectical development of the mediation of these two. Finally, philosophy itself is inseparable from the history of philosophy, which becomes the substratum for all actualizations of philosophy. In such a way, if we say philosophy is on the one hand unified out of opposition with empirical science and on the other hand bound up with the history of philosophy, it would seem natural that contemporary physics must also be mediated philosophically with classical philosophy. The intention of this essay has thus been an attempt at such mediation.

Questions for further discussion

In the final sentences of the essay, Tanabe emphasizes that his project is a 'mediation', which differs from Nishida, who focuses on the dialectical situation of the self at a site of engagement with the world. Tanabe writes, echoing Aristotle, that 'form and matter' are 'concretely understood with the exactitude of seeing ontologically in itself', but specifies that it is only an analogy for a 'reciprocal relationship' formed by 'philosophy and the natural sciences'. The words for form and matter here invoke the classical Aristotelian terms, though recontextualized in light of contemporary developments in physics, whereby matter itself is problematized. He argues that 'philosophy itself is inseparable from the history of philosophy' and thus mediation through philosophy 'is on the one hand unified out of opposition with empirical science and on the other hand bound up with the history of philosophy' so that what is known epistemologically and ontologically through observation is bound up with the history of thought and the engagement of science.

The importance of this point in understanding differences in the philosophical methods of Tanabe and Nishida cannot be overemphasized. Here Tanabe is grounding his philosophy of the dialectical mediation of *fields* of investigation in a rigorously interdisciplinary approach that *as* mediation in his framing itself situates the subject–object dialectic as peripheral to the thickening of the terms and discourses of mediation themselves. Nishida tends throughout his writings to (1) focus on the subject itself and (2) situate the relation of his philosophy as a clearly intertextual and interdiscursive relation of engagement of philosophy with philosophy in a broad sense of including writers in psychology and physics who are already writing in a humanistic way. The interdisciplinarity in Nishida is one of borrowing metaphors *to develop his system*. Interdisciplinarity in Tanabe opens avenues of otherness to enter by way of a recognition of evolving forms of recognizable being due to the sciences and not merely due to the development of the discursive forms alone. Simply put, Nishida is closer to Wittgenstein than to Derrida in terms of the intertextual care by which he constructs his arguments that are in the end clearly rooted in metaphysical dialectics that are based on anthropocentric systems. Tanabe offers a vision that while not post-human is clearly capable of being recognized as paving the way for an integration of science and thought; however, first one must engage, as some have, with how Tanabe's wartime thought, the logic of the species or the specific, can be understood so as to excise or bracket its apparent racist implications without simply circumventing lingering issues in this postcolonial minefield. If this can be accomplished, how does Tanabe suggest not an overcoming of the West but rather of the 'formalism of critical philosophy', as he says, by way of science? Then, does this amount to some form of scientism (privileging of science) or a complication of human placement in the world that may be said to constitute a proto-post-humanism?

Tanabe writes that 'without a doubt the only appropriate way to interpret the relationship of macroscopic continuous matter and microscopic discontinuous atoms is dialectically. This must include the relationship of waves and particles according to today's new quantum theory.' He introduces to these debates the phrase 'dialectical ontology' so as to crystallize for us how though there is no Kyoto School by design, there are characteristics that approach the status of motifs or methodological assumptions that define the Kyoto School turn from what are givens in so-called Western philosophy to a critique, notably by way of placing nothingness (*sunyata*; *mu*) as a natural starting point for a Japanese or Asian engagement with this philosophy. As such, it has the potential to initiate a discourse (even today) that might integrate it and Japanese philosophy together with world philosophy (see Chapters 1 and 5). Moreover, by situating physicists in this discussion he recognizes several aspects of the discourse he engages: (1) physicists are important to the realization of physics as an investigation of matter and complications that form relations between observing subjects and matter as mediated by formulae, including ones invoking problems in vector matrices and probability for waves and particles in various system states; (2) the philosophy of physics is primarily rooted in discourses on ontology, epistemology and phenomenology, so that physics is used to invoke exemplars that *potentially* alter how we understand the world; and (3) it is indeed a 'dialectical ontology' that Nishida, Tanabe and even their Marxist students such as Tosaka brought to their respective writings, though each in unique ways.

Another key passage in the closing paragraph of this text is worth noting:

> Ontologies of the classical age put forth categories for astronomical objects certainly so as to render an ontology of natural existence, likely still significant in ontological form in relation to physics today. On the affirmative side, this nature is even as a living nature remarkable in an Aristotelian way, while on its negative side, in terms of the substratum of matter it corresponds to physical nature. Consequently, there seems nothing particularly odd even though one claims that this ontology takes a form in relation to physics. In this regard, I think to attack interpretations of classical ontologies by means of concepts found in the new physics as anachronistic would be to draw a conclusion without understanding the inseparability [*sōsoku*] of ontology and empirical cognition. On the contrary, to contemplate and philosophically mediate this inseparability, whether for classical ontological interpretations or critical contemporary physics, would in either case seem to carry significance. From critical philosophy's standard view of the relationship

of philosophy with the natural sciences, one simply sees from the angle of the latter's discovery of transcendental foundations for conceptual configurations. From this ontological viewpoint one sees a unity of the opposition of phenomenal reality and an essentialist structure of existence that is considered to be of the highest importance today in order to avoid the dogmatism of the old natural sciences and at the same time overcome the formalism of critical philosophy.

Here the Kyoto School motif of a dialectics is biased not toward the Hegelian construction of being but rather a general Buddhistic void (*Śūnyatā* or *mu*, the void of undifferentiated existence). Thus the alternation of affirmative and negative is both local, to the dialectic described in the perception of nature and by way of the implicit intertexts and frames invoked, and global with respect to dialogues with multiple groundings. Also, 'substratum of matter' invokes an Aristotelian model of matter. Here in the concluding chapter of the book, we can see Tanabe invoke a term that supports the particular approaches revolving around dialectics as a necessary joining of what is inseparable – 相即 – a term borrowed from Buddhist thought and developed in this way in the Kyoto School.

Finally, one can find a perfect mirroring of both Nishida's dialectics of mutual negation as emergence of being as situated in physics by Tanabe, who writes:

> In comparison with the appearance of subjectivity in the theory of physics, one should say that it presents the self-negating structure of physical existence itself. When you do so, so-called observation is not opposed to the objective as a subjective act, but might be termed the moment of the dialectical self-realization of existence itself.

This 'self-negating structure of physical existence itself' articulates a dialectical union of investigations in physics and philosophy, specifically issues of resituating philosophy in the Kyoto School in light of the impact of modern physics. This point is of interest in multiple senses, including not only the convergence with Tosaka's historicization of physics as a field and Nishida's focus on the material engagements of experimentation in physics, but also the very broad turn here away from theoretical physics and toward practical applications, which indeed characterizes quantum mechanics, insofar as (outside of the philosophy of science) it has pushed such questions to the side while getting on with applications of it.

1. Tanabe is certainly challenging to read, but what drama is playing out in this work? What narrative seems to be central. Is Tanabe or Dirac the hero, in this sense?
2. Initiate a multidisciplinary discussion of the text in class with colleagues from physics, East Asian sciences, etcetera. Do they draw out other aspects of the text?

3. Are there some theories which are incompatible with modern physics? Try to understand why.
4. When Tanabe uses yin and yang to understand electrons and positrons, what wanted and unwanted connotations might be said to be implied? Yin is associated with the moon, the ground (Earth), lower positions, darkness, stillness, femininity, poverty and being unlucky. Yang is associated with the sun, the sky (heavens), upper positions, light, movement, masculinity, wealth and being lucky.
5. Having read this work, do you come away with more of a loss of faith in science or a deepening interest in it?
6. How might we expand on Tanabe's work and map questions of scale onto quantum phenomena?
7. How does Tanabe map the language of physics onto key concepts in Nishida Philosophy found in Chapter 2?
8. Does Tanabe map quantum phenomena as emergent from an undifferentiated void?
9. How does the new physics empower Kyoto School philosophy through Tanabe?
10. How might a 'hole' have dramatized ontological emergence for Tanabe (when he first encountered Dirac's work)?

References

Barad, Karen, *Meeting the Universe Halfway: Quantum Physics and the Entanglement of Matter and Meaning*, Durham, NC: Duke University Press, 2007.

Faye, Jan and Henry J. Folse, *Niels Bohr and the Philosophy of Physics: Twenty-first-century Perspectives*, London: Bloomsbury Academic, 2017.

Mine, Hideki, *Nishida tetsugaku to Tanabe tetsugaku no taiketsu: basho no ronri to benshōhō*, Kyōto-shi: Mineruva Shobo, 2012.

Studtmann, Paul, 'Aristotle's Categories', in Edward N. Zalta (ed.), *The Stanford Encyclopedia of Philosophy* (Fall 2018 Edition), https://plato.stanford.edu/archives/fall2018/entries/aristotle-categories/.

Wakichi Miyamoto et al. (eds), *Iwanami tetsugaku jiten*, Tokyo: Iwanami Shoten, 1922.

4
Modern Physics, Space and Ideology in Tosaka Jun

Introduction

Focusing on the outspoken cultural critic Tosaka Jun (1900–42), this chapter introduces his key writings on space, enmeshed as they are in debates in the philosophy of science, the role of space in conceptualizing social praxis and discourse, and how modern science presented challenges to doctrinaire Marxists. Tosaka, along with Miki Kiyoshi (1897–1945), reflected the culminating influence of Marxism in the 1920s while studying under Nishida and Tanabe at Kyoto University. They each chose to adhere to their political convictions after many others had given into government censorship pressures – backed by a 'Public Security Preservation Law' that presented communism as a threat to the national polity (*kokutai*) – and both ended up in prison and dying young. Though attacked as bourgeois idealists by Tosaka, Nishida and Tanabe influenced him and he them, as they can be seen to have variously made accommodations with historical contextualization, reflecting the impact of Marxist journals that each edited.[1] While a charismatic figure, Tosaka was certainly, compared to Nishida and Tanabe, the most limited in his comprehension of the finer points in quantum physics, apparently constrained by the Soviet party line on matters of science in his later work. Yet, if one reads his various writings on space, the sciences and physics in particular, one cannot help but be impressed by his vast encyclopaedic knowledge and flair for a dynamic, dramatic style that depends more on name-dropping and quick summaries of keywords than on sustained exposition. Thus, the first impression one has of him upon glancing at the texts and the final impression after reading them may is one of anticlimax, which is certainly not the case when reading Nishida and Tanabe. Thus short excerpts that represent his own views are included, though there are many other passages that might have been chosen if the purpose had been merely to mirror his own high-energy, low-yield arguments.

As Tomio Nishikawa has pointed out, 'for Tosaka materialism functioned as an ideology of "critique"' and was the means of 'radical criticism ... that could be used to penetrate every possible field'.[2] Thus, while Nishida neatly integrates an intertextual assemblage from disparate philosophical sources, Tosaka repels such sources, sometimes only seeming to sustain his discourse through an overbearing voice in his prose, one that substitutes for reason and analysis. Research on Tosaka in English tends to treat him as a martyr, which he certainly was from the point of view of progressive politics; he was a critic capable of forging new methodologies for addressing problems with liberalism that remain provocative and ahead of their time. Nevertheless, from the perspective of his writings on quantum physics, it is quite clear that he has not bothered to update himself on these topics so similar to ones he assiduously pursued a decade earlier, before the field became complicated.[3] Tomio Nishikawa characterizes him as 'by no means a materialist and a Marxist in the sense of someone involved in the revolutionary movement. What gave Tosaka his distinctive character was that, as a critical rationalist, he provided a comprehensively radical critical investigation of ideology, including natural science, social science, and culture.'[4] However, one should also note that in his writings on physics, Tosaka sides with official Soviet interpretations of quantum physics, seemingly to avoid confronting the conundrums it presented. Indeed, his journal *Research in Materialism* is modelled after Soviet journals.[5] This issue will remain significant throughout this chapter.

Tosaka, like Nishida and Tanabe, was torn between studying mathematics and philosophy. Thus his early work approaches space primarily with the natural sciences in mind. It is important to point out that throughout his later materialist writings he constantly seeks out polemics to present dramatically as contraries in dialectical relationships. Then, in standing back to decide what it all means, he usually will proclaim that a bourgeois distortion has blinded the world from some underlying truth. The truth, with regard to space, for him, is manifest in the everyday. In 'On Space' (*Kūkanron*), in *On Contemporary Materialism* (*Gendai yuibutsuron*, 1936),[6] Tosaka writes, 'In other words, space encompasses all our practices. We organize not only the actual world, of sight, touch, and sound in the determination of space, but our day-to-day life completely follows in accord with this organization of space.'[7] What he means by this is not only a joining of space-times in terms of relativity, but what amounts to a rendering of space as a term for dialectical synthesis itself. How he achieved this role for space can be located in his interdisciplinary approach to it. Understanding his method as such also suggests a complication of charges that Tosaka's dialectical approach was wanting. As Nishikawa Tomio writes, 'To Tosaka, rationalism referred to a clear-cut way of thinking that says "that is that and this is this" – the thinking of ratio, or logic. To "critique" is presumably to thoroughly engage in this clear-cut way of thinking or "thinking that divides". In this sense as well he was not a dialectician.'[8] But, in light of his division of space into interdisciplinary fields (*psychology*, *geometry* and *physics*) that variously come together through opposition, he also allowed for a non-specialist space to serve as a common ground for *everyday space* even in asserting dialectics. These three divisions of types of space give rise to various disciplines by which 'we all possess a common sense in terms of everyday life, and are no longer specialized – but are bound up as one in the everyday, commonsensical conceptions of space'.[9] In emphasizing the *thereness-nature* (*Da-seikaku*, after Heidegger) in space as a locus for the everyday, Tosaka situates the ontic – the *thrownness* of being in the sense of *Dasein* (being there) – in relation to the ontology of the material world in physics.[10] One may note that while this appears to be yet another instance of Heidegger influencing this generation of Japanese philosophers, Tosaka reflects the dynamics cultivated in the context of the Kyoto School, following Nishida's innovations, especially the application of a dialectics that begins in a void rather than with assumptions of being. Thus one can understand Tosaka, when he writes, 'The truth of space, the true determination of space, lies solely in the synthesis of this indirect form [based on the conceptual structure of specialized discourses] with this direct form [based on the conceptual abstraction of space itself].'[11] It is both mediated and directly perceived, interdisciplinary and everyday.

Though Tosaka attacks Nishida on many levels, including for being psychological and offering only a static system, Nishida still figures in Tosaka's methodology and practice insofar as it too is focused on expanding a comprehension of a subject actively engaged in the spatio-temporally dynamic world. Nishida's mediating subject relates to the 'world itself' but later also situates others, suggesting a pluralism (probably inspired by American philosopher William James (1842–1910), who devoted a book to the topic) that in the end is far more amendable to democratic institutions than Tosaka. Tosaka, for all his critical dismissals and innovative but predictable Marxist positions, offers primarily (1) a bold and highly original critique of liberalism in *On Japanese Ideology* (1936) and (2) a means for official Soviet (Communist Party) positions to be presented in Japan in a favourable light during a time when the censorship laws demanded one's renunciation of communist affiliations (*tenkō*). The second item is often omitted from accounts, though it is obvious to anyone who has examined his writings on modern science: he maintains party positions that grossly oversimplify extremely complex issues in modern physics, issues which Nishida and Tanabe did not hesitate to engage more subtly. As such, Tosaka never fully understood or allowed himself the capitalist indulgence in, he would say, the fundamental issues of how subject and object meet in quantum phenomena as understood from the mid-1920s. He dismissed such work as a ruse of 'bourgeois academia' that in effect, for him, exaggerated a so-called 'crisis' in modern physics. Thus, while long interested in related issues in space-time and post-Euclidean geometry, he never understood quantum physics (or publicly allowed himself to speak out on it). Whether he simply never bothered to try (or lacked time or resources) or because he chose to adhere to Soviet party positions is a matter for debate.

'Scientific', for Tosaka, appears to conflate a vanguard of Marxist truth – as a deconstruction of socio-economic structural capitalist dependencies on exploitation, alienation and so on – with a vanguard of truth in the natural sciences. Tosaka never seems to have appreciated the irony that with the loss of a totality and even causality inherent to classical physics, relativity and quantum physics also suggested new forms of subject–object relationality that like reflexive postwar poststructuralist thought, would lead to an awareness of the non-commensurability of base and superstructure (what now is referred to as 'vulgar Marxism'). On the contrary, Tosaka, still adheres to a vision of total transformation of socio-economic and political conditions in Japan, so that the primary point of reference is utopian (a non-place) that contrasts with Nishida's focus on places of mediation and Tanabe's somewhat Heideggerian focus on communities of mediation that reflect categories and given situations.

4.1 From Kantian to Marxist approaches to space and matter

Introduction

In one of his earliest published works, before becoming a Marxist, Tosaka wrote extensively on the various approaches to space and time first explored as a student. Focusing less on the new physics than on Kantian problems concerning the foundational or schematic role of intuitions of 'space' in the situating of secondary discourses, he takes certain spatial relations for granted. After all, Kant situated space and time as primary framing antimonies in his (for lack of a better word, 'proto-dialectical') situating of intuition of phenomena in terms of *a priori* transcendental categories that define objects within an idealist (post-Aristotelian) recasting of form and matter. Relativity and quantum mechanics would variously demand the dissolution of such *a priori* givens and the physical assumption of a universal aether, so that absolute space and time fall apart. The edifice of scientific positivism, in effect, turns on itself. In fact, it would be soon after Tosaka's publication of 'About Space' (*Kūkan ni tsuite*, 1924) that Heisenberg, Schrödinger, Bohr and then Dirac would enter into a dispute also including Einstein concerning the particle-wave nature of matter, a dispute that is yet to be resolved. Tosaka's early (and more carefully researched) writings on space and physics in general preceded this watershed event in physics, while his later work is by comparison cursory in its treatment of quantum physics. Questions we need to explore in Tosaka include how variously did he grapple with the loss of aether; and, as a materialist, how did his foundational Marxist concepts become inflected in his objective orientation, so as to inhibit him from appreciating the greater implications of quantum mechanics.

Tosaka, the problem of criticism and the production of space

While in Tosaka's early work he is interested in how space and time are recognized, represented and defined in Kantian terms, later as a Marxist he becomes interested in how such interdisciplinary delineations enter the larger social picture of the contours of power and organization of state and private interests and how ideologies are cultivated so as to justify them. Nevertheless, one can see some continuity in how he develops an understanding of space in light of competing discourses on it even in his Marxist period. Given Tosaka's long-standing interest in space, for him it is made central to ideological questions that encompass all aspects of representation within a state. The new physics complicates the givenness of space and time, so that Tosaka will be shown to take positions aligned with Soviet official positions, which he justifies by various arguments that readers may compare with positions taken by Nishida and Tanabe on similar issues.

Box 4.1: Space in Kant

Space in Kant is situated as something simultaneous manifesting itself with intuitions, yet also making them possible, as it is a concept and not a thing, per se. Kant writes in the *Critique of Pure Reason*:

> ... the representation of space cannot be obtained from the relations of outer appearance through experience, but this outer experience is itself first possible only through this representation [of space].
>
> Space is a necessary representation, *a priori*, which is the ground of all outer intuitions. One can never represent that there is no space, although one can very well think that there are no objects to be encountered in it. It is therefore to be regarded as a condition of the possibility of appearances, not as a determination dependent on them, it is an *a priori* representation that necessarily grounds outer appearances.

(See Immanuel Kant, Paul Guyer and Allen W. Wood, *Critique of Pure Reason*, New York: Cambridge University Press, 1998, 157–8.)

Given that Tosaka was interested in space both in his student work and as a Marxist, one might expect some sort of critical convergence with the work of a later Marxist exploring matters of space, Henri Lefebvre.[12] Lefebvre points to an important aspect of space and representation in this context:

> The area where ideology and knowledge are barely distinguishable is subsumed under the broader notion of *representation*, which thus supplants the concept of ideology and becomes a serviceable [operational] tool for the analysis of spaces, as of those societies which have given rise to them and recognized themselves in them.[13]

In lieu of such emphasis on the representation of spaces, within a fusion of ideological and epistemological space, the early Tosaka introduces to such a discourse on space a Kantian preoccupation with spatial issues, rendering space more definitive of ideological structures and mechanisms of intuition than the representation *of* space itself. As in Kant, representation follows from space, not space from representation. As will become clear, for Tosaka the relation of material and methodological spaces is an issue which supersedes Lefebvre's broad divisions of mental, physical and social spaces.[14] Lefebvre compartmentalizes spaces in accordance with given discourses: we are 'confronted by an indefinite multitude of spaces, each one piled upon, or perhaps contained within, the next: geographical, economic, demographic, sociological, ecological, political, commercial, national, continental, global'.[15] Tosaka, like Kant, seeks a dynamic critical account of relations that insist on what may be considered interdisciplinary, but not in a sense of founded on discrete disciplines per se. For Tosaka we can never grasp the nature of space unless we examine the relationship of totalities to parts in a more dynamic way than a to-and-fro between temporality and spatiality. Tosaka's early work traces a genealogy of the domination of space *concomitant* with the rise of the enlightenment in light of the assumed conjunction of absolute space and Euclidean geometry, and how these have been de-absolutized (not negated altogether) by relativity theories.[16] There is a renewal of local and non-absolute measuring subjects within open space, so that Tosaka argues axioms do not produce intuitions but intuitions axioms.[17] In short, the empirical positivism of scientific method supersedes metaphysical systemic assumptions here.

But Tosaka does not simply subordinate productions of 'science' to critical themes and discourses laterally associated in a literary rhetoric.[18] Tosaka's early use of 'space' certainly makes possible an original resituating of temporal emphases on a Marxian critical authenticity associated with immediacy, for instance, building on the work of Nishida and many others. It becomes clear that his later ideological critique not only resists Japanist thought and other visible or tacit support of right-wing Japanese nationalism (which he identifies with liberalism itself). Whereas Nishida exemplified a stylistic tour de force integrating multiple sources left largely intact, though recast within his void-based dialectics, Tosaka may be seen to focus on the transition from *a priori* systematic metaphysics to a multiplicity of recognizable spaces and spatialities. He does develop a place for space in his original critique when arguing that space itself must also be analyzed, and not only by analogy: 'space as analogy must not be confused with common sense conceptions of space'.[19] However, his later thought privileges common sense in this debate over space.

It would appear in this regard that Tosaka was most influenced by the work of German philosopher Oscar Becker (1889–1964), who went beyond Kant's transcendental critique that assumed a given frame by which space and time were invoked carte blanche as vacancies to be filled. Becker, Tosaka outlines, distinguishes three types of representation of space: prespatial sensory (visual, tactile, auditory) fields indicated by movements in our sense organs; orientative space that maintains a depth established in the relationship between the self and the external world; and through repetition of the singular point of the self, a representational homogeneous space. Intuition in a post-Kantian sense is the third type, self-centred, as it were, since relativity no longer affords the certainty of Euclidean or absolute space, thus undermining a key component of Kant's transcendental critique. Becker, Tosaka writes, saw Kantian intuition of space within a Euclidean frame now as a 'phantom-space' (*Phantomraum*).[20]

It is by way of a Marxist critique that Tosaka attempts to fulfil not only a Marxist ideal, but the influential Meiji educator and writer Fukuzawa Yukichi's (1835–1901) vision of making Japanese intellectuals the watchdogs of state apparatuses.[21] His work arises in an indefatigable discursive space of disclosed contradictions and overturned assumptions that he challenges to be newly justified. Thus, while Nishida and Tanabe constructed systems, Tosaka turns this void-orientated dialectics into a method for situating social criticism. Following Marx, Tosaka is concerned with the very encroachment of capitalistic forms of thought and culture that ideologically disarm counter-actions and counter-narratives by naturalizing their exteriority according to the sublation of an 'unconscious' social *common sense*. Common sense plays an important role in Tosaka's thought, and yet it is by its very nature general. It is *as if* it were without a defining, determinable origination, but existed outside of history. For Tosaka, it is something which would seem to elude one, being an underlying starting point constituting the atmosphere of the social itself. For him, common sense must be examined in terms of what it entails, what it means, and what beliefs and assumptions have become actively reproduced *without resistance*.

Tosaka's task can be seen as the opening of a critical space which defines the basis for a critique of thought which attends *not only* to the appearance of opposition, conflict, contestation and all the elements of change that define history most generally, but the production of space and space-time themselves (which immediately seems a post-Kantian interest in the framing of space and community). It is here that one can see how appropriate Tosaka's use of 'space' is, along Kantian lines, as an 'archetypal' spatialization, a problem of relation merging theory and practice, reason and understanding, discourse and intuition, representation and presentation. He is interested in revealing the points in the present discursive activities which provide not only resistance to immediate change, but absence of resistance: the non-contested reproduction of an idea, its genealogically discernible reification, for Tosaka, defines a crucial aspect of the modern thought. Kant becomes a starting point in his study of space and a recurrent point of reference in his analysis of the ideological. As will become evident, *space* is crucially bound up in Tosaka's critique; space attains a physicality which goes beyond the absoluteness that enshrouds, defers and defines Kant's thing-in-itself (always unknowable). So as to resist the mental of phenomenology and hermeneutics, Tosaka posited the importance of *physical time* along with physical space. As we today still struggle with questions of the place of a moral or critical high ground in modern or postmodern societies, the issues of space as treated by Tosaka may be helpful in determining how historical and genealogical analyses are variously situated in this debate today, both within it and outside it, in a technology of inside/outside, of limits – without elevating a 'threshold' or 'limit' to a definitive status within a logic. It is here that what may be termed his *discursive* intertextuality (not primary or semantic), or what Marc Angenot calls interdiscursivity, comes into focus. As such, while Tosaka is in principle a card-carrying Marxist in practice, he is a liberal in his openness to established common sense even against the evidence of quantum mechanics, which he treats almost as a conspiracy theory of the bourgeoisie, who destabilize common sense for him. He provides an analysis which inhabits the space it analyses, superficially, not unlike Nishida; however, unlike other Marxist critiques (which tend to situate their own critiques themselves as reflections of the political, economic and cultural climate) he allows for and defends a critical space which is itself determined within a dialectic of totalities and constitutive aspects.[22] Is it not this broadly Nishida-styled emplacement of the subject in 'history' that Tosaka both inherited from Nishida and brought out in Nishida as part of a larger materialist critique of it?

Space, following Kant, comes to have an empirical reality insofar as things by themselves are dependent on it, but it is also manifest as a pure intuition, *a priori*, that is apodictic, universal and necessary; however, 'space refers to the pure form of intuition only, and involves no kind of sensation, nothing empirical'.[23] Here too, Nishida's influence figures prominently. Tosaka, in his critique of ideology, returns to and develops Kantian issues in his relation to discrete spheres of discursive attention. Where there is an understanding of interdependencies between phenomenal processes of intuition and the space of intuition, Tosaka treats various (ideologically charged) discourses as embodying mechanisms which constitute *possibilities* for the various discourses operating at a given time. These mechanisms are analogous to Kant's 'forms of intuition', though a major difference must be emphasized: for Kant, thought can be hierarchized from simple intuition to synthetical judgement, from primary to secondary discourses. In Tosaka's critique of Japanese ideology, liberal thought comes to engender mechanisms, hidden yet definitive principles not immediately apparent in its espoused position, within a given discourse put under scrutiny. This is not to say that Tosaka is interested in epistemology per se in the sense that philosophical idealism portrays itself, as a metaphysical search for reductive epiphanies; rather, Tosaka focuses on discursive productions as secondary discourses, and how within them an 'order of meaning' has supplanted an 'order of things'.[24]

Tosaka's use of Kant suggests a paradigm helpful in understanding how he generates a genealogical approach to the interdiscursive relations. To interpret ideology requires a grappling with the relation of types of logic or reason (*ronri*); interests and control in academic, journalistic and everyday discourses; and the production of appearances and ironic relations within hegemonic structures which naturalize and rationalize the dispersion of power. This last aspect involves ironies of incongruities between the mechanisms defined by discursive parameters and the effects manifested within such public, governmental or academic systems.

Thus, for instance, the appearance of liberal discourse involves not immediately apparent interests and reasons for maintaining given positions; the reasons need not relate to the manifest discourse in a causal manner (mechanistically, from inception to implementation, principle to product), nor an organic manner (from seeds of ideas to branching systems). The underlying principles themselves are in Tosaka, following Kant's paradigm, *a priori* yet necessary for the manifestation of a given discursive space; it is genealogically related to the discursive space, so as to invite the probing of contemporary discourses, in relation to each other and past discourses, in an intertextual relationship influenced by a Kantian interpenetration of space and knowledge. In Tosaka's fundamental, early interpretation of Kantian intuition-space, he writes:

> [T]here must be a given synthesis behind the unity of... representation. What is clear about this synthesis is that it is without doubt an 'axiom of intuition'. In order for each and every phenomenon to be incorporated in experiential consciousness (in the sense of a factual consciousness) they must without exception depend upon a synthesis of *das Gleichartige* [the homogeneous] and a consciousness of the synthetic unity of a multitude of similar things. At the same time, 'if there is not even the smallest of lines drawn from within our minds *in Gedanken zu ziehen* [drawn from thoughts], namely, if one cannot initially indicate the intuition of a "line" following the successive production of every segment from a point, we can have no representation.' In this way the representation of the parts becomes the basis for the representation of the whole, and that which precedes it is called the extensive quantity. Space is thus nothing other than an extensive quantity. In accordance with 'principles' of the axioms of intuition, 'all intuitions are extensive quantities.' An intuition which is unified as an extensive quantity is, as outlined above, made possible by an orderly synthesis of parts.[25]

The question is, how could the same person who wrote the above then misunderstand the fundamental problems of quantum mechanics, namely, its demand for an account of observation in quantum physics, whereby situated space is reduced to minimal subatomic systems? Such 'extensive quantities' are what Tosaka will translate, in a theory of ideology, into discursive possibilities and limitations. He continues:

The intuition of space unified as an extensive quantity, seen from its objectified unity, must be an objectified formal intuition. If so, what is the orderly synthesis of parts which can be thought to be based on this objectified formal intuition? For Kant, the orderly synthesis of parts in accord with a consciousness of space is none other than the *produktive Einbildungskraft* [productive power of the imagination]. That which brings about a consciousness of reality in general is actually this productive imagination. The productive power of the imagination is this orderly synthesis of the legal itself, which establishes the intuition of space. These laws are the unification of representations of space itself. Thus, as engendered in accordance with these laws, the *Produkt*, namely, that which itself resembles a figure taken as a 'schemata' [*zushiki*], is this objectified formal intuition. This synthesis of an order of a productive imagination must itself be exactly appropriate for formal intuitions which as of yet have not been objectified. We must go beyond considerations of space as simply something objective, and reflect on the concrete formal intuitions of space. In formal intuitions in the imagination, the opposition of objectified and not yet objectified intuitions are directly linked. The reason that space is an intuition appears to Kant to be contingent upon its appearance as a formal intuition. Thus, though formal intuitions are already interpreted by way of the imagination, for Kant, we must expect the imagination to be bound up with given categories. This means that the laws of the orderly synthesis of the imagination follow categories. In other words, formal intuitions may to this extent be called categorical. Nevertheless, as is said, 'categories' are none other than norms of recognition. Formal intuitions must be normative. The *a priori* that Kant recognized in space is actually this norm. *Intuition-space is the norm*.[26]

Intuition-space becomes in Tosaka a means of accounting for the moral or political implication of 'actions' without confining oneself to either the positivities of causal explanations, which unconsciously organize social and individual psyches. But how does this confine his thought later to a rather limited understanding of issues in quantum physics? (See also Chapter 5 for further investigation of quantum mechanical forms of intuition in contemporary discussions of the philosophy of science and physics.)

Kant and Einstein: matters of movement and the dialectics of matter

Tosaka rescues Kant from his Newtonian milieu, by which the absolute in general as a foundation for knowledge depends on the absolute nature of space. Instead, he recasts Kantian problems concerning the relation of what is *a priori* to contextualized knowledge-production and indeed space-production following Einstein, whose relativity theory he in effect converts into a Marxian approach to space whereby physics is understood less as a pure theoretical science than as a materialist discourse that challenges bourgeois idealist assumptions. In 'Ideology and the Natural Sciences' (*Shizen kagaku to ideorogī*), in *Lectures on Contemporary Materialism* (*Gendai yuibutsuron kōwa*, 1934), Tosaka writes:

> Thus the interior moment of the natural sciences – its logical, methodological structure, and its fundamental concepts – and its exterior moment – the historical and social existence of the natural sciences – can be initially synthesized for the sake of working through [*baikai*] the existence of an ideology. What is called a crisis in modern physics – and the natural sciences in general – is actually nothing other than a crisis in bourgeois ideology (in its mechanisms, metaphysics). Thus, this collapse of ideology – originally riddled with shortcomings – which is necessarily perceived as some sort of crisis, renders in our consciousness the process of collapse in capitalism as a crisis in society itself. Insofar as the process of collapse in capitalism is identified in our minds with the progress of society, there is no crisis in modern physics, but only rapid progress in physics. This is the essence of the [purported] crisis in modern physics.[27]

In short, Tosaka points out how bourgeois ideology, in its straining towards a fiction of social, personal and economic 'equilibrium', tends to project and find its own sense of immanent crisis in physics. This relation of physics and philosophy he points to demonstrates the rigidity of liberalism in a way in which Tosaka is quite adept. For all the rhetoric of liberal 'openness' to the new, unique and different, as a rule relativity theory is confined to the margins as a disruption of the established ways of understanding our world. It is here that the primacy of 'space' becomes doubly clear, and the aptness of Tosaka's critique establishes its validity: relativity theory, as he uses it, we may say, accounts for its critical vantage by resituating a tradition of metaphysics and idealism, rooted in Newtonian physics, in light of Einsteinian physics. He asks, what changes in the foundations of our epistemological methods are potentiated by this progress in physics?

In 'On Space', he builds his understanding of 'space' in terms of three categories of space generally recognized: psychological, concerned with representation; the geometrical, related to mathematics; and the physical, concerned with substance.[28] Though these three divisions of types of space give rise to various disciplines, such as psychology and physics, 'they all possess a common sense in terms of everyday life, and are not ipso facto specialized, but are bound up as a unity in everyday, common sense conceptions of space'. 'The truth of space, the true determination of space, lies without exception in the synthesis of this indirect form and this direct form.'[29]

Space is both mediated and directly perceived. The ultimate dialectic with regard to space is that it is physical, yet geometrical. Of the three disciplinary treatments of space – intuitive (psychological), geometric and physical – there is one that stands outside of 'the knowledge of specialized sciences', this is 'non-specialized, everyday space'. If these three spaces are '*partial* forms of the phenomenon of space itself', everyday space is an '*overall* [*zenpanteki*] form of the phenomenon of space itself'. 'We may say that, if the former is *indirectly* an abstraction of space itself, everyday space is a direct abstraction of space itself.'³⁰ But, seemingly paradoxically, criticism constituted in terms of everyday space falls under the rubric of philosophy, 'for philosophy, which arises out of an *unific* interest, the many concepts of space make this everyday space central and returns to this area'.³¹

For Tosaka in this period, the concept (*gainen*) of everyday space should not be confused with the everyday per se:

> Kant's analytic method does not rise above *psychological* and *phenomenal categories*. Instead of recognizing the analysis … of space *by way of everyday concepts*, it merely analyzes space *as an intuition*. We analyze (in what we shall provisionally call concept analysis [*gainen bunseki*]) space *as something in terms of everyday concepts*, and also as *an everyday concept, in terms of everyday concepts* as well as *everyday ideas* [*kannen*].³²

Thus Tosaka in the end reverses Kant's assertion that 'the *total system of idealism* is motivated by the *discovery of the idealism of space*'.³³ Instead, materialism, not idealism, should be recognized as most suited to the delineation of problems of space.

Translation

from 'The Scientific World' (*Kagakuteki sekai* 科学的世界), Chapter 6, *On Science* (*Kagaku-ron* 科学論, 1935)³⁴

Matter that has been concretized and particularized by categories in physics becomes *things*. Things may be considered ordinary *physical objects* [*buttai*]. According to atomic theory, they are *minute particles* [*biryūshi*]. Even so, according to the wave mechanics established by de Broglie and Schrödinger, matter may be considered *matter waves* that are particular combinations of a kind of wave. In relation to light, the determination of particles [photons] by Newton and the determination of waves by Huygens have already stood opposed. Yet light, along with electromagnetic waves and ordinary radiation, has until recently been understood by way of determinations of waves. Nevertheless, after the discovery by Planck that energy in general has a species of minute particle units called *quanta*, Einstein then clarified the existence of *light particles*. Thereupon, light and matter are generally equivalent with energy, and must be specified by the same determinations. As such, on the contrary, the wavelike nature of light even comes to be brought to bear on matter. Thus it is said that matter is both wave and particle, mutually exclusive determinations that have no choice but to remain unified and, historically speaking, reciprocally contradictory.* To put it in general terms, in proving the dialectical unity of the *intermittent* and the *continual*, what leads mathematically from the intermittent to the continual (infinite as well as **transfinite**) is **Cantor's set theory**; however, today two of its approaches appear opposed: **Brouwer's** appeal to intuitive continuity in **intuitionism** and **Hilbert's** appeal to axiomatic mechanisms in formalism. Though what should sublate the contradictory opposition of the mystical theory of intuitionism and the mechanical theories of formalism should indeed be a dialectic, the philosophy of mathematics has yet to reach such a stage as to prove one.

On the other hand, matter is said to have lost its relation with

> Box 4.2: Christiaan Huygens (1629–95), Georg Cantor (1845–1918), Luitzen Egbertus Jan Brouwer (1881–1966) and David Hilbert (1862–1943)
>
> Astronomer and mathematician **Christiaan Huygens** is known as the originator of the wave theory of light and discoverer of the rings of Saturn. **Georg Cantor** is best known for his theory of **transfinite numbers**, which distinguish numbers of a set considered infinite as expressed in subsets, so that infinity becomes more complicated than just an endlessly postponed endpoint. **Luitzen Egbertus Jan Brouwer** promoted **intuitionism** in mathematics in response to **David Hilbert**, whose formalist approach was dominant.

space and disappeared. According to relativity theory, the potential of matter, gravity, electromagnetism and the like all are inflected by various types of bending, contraction and expansion in the *space* of the universe. Yet, actually this space [physical and mechanical space] differs from simple geometrical space. Actually it itself has *material contents* [*busshitsuteki na naiyō*]. The material concept called the aether was the first to be superseded. A *field* of forces [*chikara no ba*] marks the significance of this space. Then, it is precisely this field that becomes the new concept of matter. Thus, according to the concept of the field, matter is unified with space. Moreover, the internal connection – the unity of opposites – of determinations of time and space in terms of relativity theory is quite renowned.**

*On this point, see Werner Heisenberg, Erwin Schrödinger, Paul Adrien Maurice Dirac, *Die Moderne Atomtheorie*, 1934. [Tosaka's note, referring to a publication produced on the occasion of their Nobel Prize.]

**My *On Space* [*Iwanami*] goes into some details concerning the analysis of the concept of space. One must draw attention to recent developments in quantum mechanics that have cast doubts in relation to spatial description in the natural sciences. For example, see N. Bohr, *Atomtheorie und Naturbeschreibung* (1931) [*Atomic Theory and the Description of Nature*, 1934]. However, in my view, the concept of space *of the future* in physics clearly suggests that it must now be changed. [Tosaka's note.]

...

However, the primary determination of matter as *motion* must now be recollected. That is to say matter has been changed, developed and transformed. This itself necessitates a situation that demands a dialectics of matter itself, a dialectics of *nature itself* [see Box 4.3]. From stellar objects to the earth, the universe includes various forms of matter and living things (moreover including human societies) in a *temporal process*. The universe, matter and nature have as their *fundamental law* this *historical movement*.[35] The historical movement of nature itself becomes the most representative case of the most fundamental nature dialectics.

Box 4.3: Nature dialectics or dialectics of nature

Nature dialectics or **dialectics of nature** form an ongoing theme appearing in countless articles in the journal Tosaka edited, *Studies in Materialism* (*Yuibutsuron kenkyū*). With very few exceptions, nature is subordinated to concerns over human social practices and history, not natural history or science, though Tosaka upholds science in the abstract as what justifies Marxist critiques of ideology and capitalism. Thus while for Tosaka Marxist materialism is scientific and rooted in nature, yet it would be more accurate to say that nature is situated within its system. One may trace such thought from works such as *Anti-Duhring* (by Karl Marx and Frederick Engels, 1877) and Engels's unfinished *Dialectics of Nature* (1883), through to Vladimir Lenin's *Materialism and Empirio-Criticism* (1908) and various debates after the Russian Revolution that interested Tosaka.

One may better understand Tosaka by linking basic premises of Marxist approaches to issues in classical physics to their source in passages such as this from *Anti-Duhring*:

> Matter without motion is just as inconceivable as motion without matter. Motion is therefore as uncreatable and indestructible as matter itself; as the older philosophy (Descartes) expressed it, the quantity of

motion existing in the world is always the same. Motion therefore cannot be created; it can only be transferred. When motion is transferred from one body to another, it may be regarded, in so far as it transfers itself, is active, as the cause of motion, in so far as the latter is transferred, is passive. We call this active motion *force*, and the passive, the *manifestation of force*. Hence it is as clear as daylight that a force is as great as its manifestation, because in fact the *same* motion takes place in both. (Frederick Engels, 'Philosophy of Nature: Cosmogony, Physics, Chemistry', in *Anti-Dühring*, 1877.)

Such thought is rooted in a world of aether, force and causality with absolute spaces treated as naturally limited. This implies a lesson regarding human exploitation based on control of motion as forms of active and passive force. This outlook would contribute to the inhibiting of Tosaka's understanding of basic problems in modern physics, which indeed undermines the very premises for such Marxist thought.

Time thus comes to have special significance. Philosophers have offered various thoughts on time, including psychological time, historical (and historiographical) time, theological time and also time in physics. Yet, it is the basic sense of the time of physics that will guide us from here. This *cosmic time* provides an *order* of all of existence (including not only nature but also the history of human society). This corresponds to the so-called spine of a nature dialectic. Incidentally, according to the natural sciences, the *various laws* of nature that appear all occur in cosmic time. The reason is that with the *fundamental laws* of nature have followed a nature dialectics in terms of the movements of the changes, developments and transformations of nature.

Yet this fundamental law (in other words, what had been the standard setting for a nature dialectics), which expresses a fundamental relationship in terms of this cosmic time, has two problems. One is *causality* and the other is *theories of the evolution* of the universe. One must agree that causality is indispensable to the internal structure of historical existence. Distinguishing historical processes from merely juxtaposed orders accords with a fixed *continuous relationship* placed before one between what comes before and after in this sense of them surely being *processes* as well as *changes*. Cosmic time continues. It is built on processes, changes and continuity, while the maintenance of the status quo, stillness and discontinuity are all possible, it is exclusively by way of cosmic time that the line of *history's* connections is drawn so that these things are possible.[36] Then, this continuous relationship (with discontinuity first established upon it) between what comes before and after is commonly referred to as *causality*. When understood within a mechanistic perspective, causality falls between determinism and fatalism. Accordingly, all things are considered to be [subject to] causality=causal necessity, which is to say *completely* specified by the *absolutely fixed* according to *mechanical necessity*. However, according to the uncertainty principle based on Heisenberg's theory of quantum mechanics, this sort of concept of metaphysical causal necessity has been proven to be no longer viable.* Causality has become none other than a *necessity* (a substantive, pragmatic moment maintaining everything from the *coincidental* to the *possible*) generally clarified by material dialectics in a single physical situation. Thus, from this point of view, the concept of causality in the natural sciences inevitably must arrive at this material dialectical understanding.**

> *See W. Heisenberg, *Die Physikalischen Prinzipien der Quantenmechanik* (1930). [Tosaka's note. English translation: *The Physical Principles of Quantum Theory*, Chicago: University of Chicago Press, 1930.]
> **When we see from this perspective, though it would be off the topic to mention that so-called *Necessitarians* try to speak ill of *materialists*, I think it is often the case. However, 'I proclaim in truth to you, today you shall be in paradise together with me.' [Tosaka's ironic note unwittingly underscores the element of faith necessary for both religion and Marxism in Tosaka's context.]

Questions for further discussion

In examining Tosaka's context, as an unrepentant Marxist after most intellectuals had renounced it (*tenkō*), he in effect flaunted his opposition to Japanese nationalist sentiment during a time of intense national militarist activities in Japan and its colonies, skirmishes with its rivals for territory and resources, and even all-out wars. The journal he edited, *Yuibutsuron kenkyū* (Studies in Materialism), published articles mostly devoted to forging a Soviet-centric worldview that on the one hand can be justified conceptually as a critique of imperialism, exploitation and alienation under capitalism. On the other hand, it seems to misunderstand the structural contradictions (or dialectics, to turn Marxism against itself) of the political economics of Soviet imperialism and coercive policies that were under Stalin far worse than anything the Japanese could dream up for its policies.

Marx saw himself as inverting Hegelian idealist dialectics that sublates processes of human engagement with the world in abstractions culminating in the spirit of a people. Engels extended this inversion to an overt dialectical materialism, a highly influential concept that codified Marxism as following a clear alignment of progress and time, justifying Marxism as a utopian science. However, its contribution to the philosophy of science is considered a historical curiosity now, arising as it did in a manifest world of Newtonian physics and Darwinian views of survival of competing species, within which social modelling of humans was mapped by analogy. Problems of subatomic or cosmological scales would appear irrelevant distractions from the social sphere and scales of attention. To the extent that Tosaka followed Soviet doctrine in adopting Engels' extension of Marx into this materialist and nature dialectics, it would seem Tosaka simply accepted science in such

Marxist terms as comprehensive in its capacity to explain both human society and nature.

Though now seeming naive, when one factors in the influence of Nishida's own dialectics of *basho* situated in undifferentiated nothing, one can better understand Tosaka's possible reasoning for buying into a 'nature dialectics' as perhaps not only following the Soviet party line but also reflecting Nishida Philosophy approaches. Though Tosaka attacked Nishida Philosophy as idealist and bourgeois, Nishida's undifferentiated nothingness is a frame that never appears as such in Soviet writings but does in Tosaka. After all, Soviet adherence to a dialectical process of negation of negation (in sublation) that keeps close to material conditions defined by contradictory unities of opposites is superficially very close to Nishida's dialectics of mutual negations of world and subject.

1. How in the context of a post-1917 world saturated with socialist and communist discourse on revolution might Tosaka have justified his support of Soviet ideology and policies regarding science and the philosophy of science in Japan?
2. How could Tosaka ignore the findings of (the new) quantum theory regarding the role of subject and object being intertwined even in acts of observation? (Perhaps refer in part to Chapter 3.)
3. How does Tosaka interpret the crisis in modern physics as a bourgeois crisis?

4.2 Tosaka's critique of the crisis in modern physics

Introduction

Tosaka sometimes becomes quite passionate and even opinionated to the point of seeming to refuse to acknowledge another perspective. When it comes to quantum physics, he tries at various times to get his head around the conundrums it incites from the perspective of classical physics, sometimes to the point where his Marxist framework demands an answer that drives him into an odd corner. Below, Tosaka argues that the definition of subject lives documented in the fiction of Dostoevsky demonstrates a benchmark for grasping 'universal human nature' while presenting an 'interior' view 'nuanced' with 'substantial differences and oppositions' in fiction; however, 'artistic value' itself he sees as opposed to 'political value'.[37] To define an exterior world of valuation, he seeks political value in not only art but 'concretization in terms of science'.[38]

Tosaka argues unconvincingly that the sciences of the interior are cultural and political while the sciences of the hard sciences have no interior cultural or political 'moment'. He calls the hard sciences based on theory – the new physics – as 'nothing but "external" contingency', that is, science is based on coincidences as 'inner moments', which become the sole means of grasping science by a theoretician.[39] Tosaka appears to have no appreciation of the centrality of communicating relationships in physics by way of symbolic means known as formulae, nor of the role of probability, matrix mechanics, wave mechanics or much else. The works he cites are primarily classical physicists or writers who can be read in a way that conforms to his political agenda. When he does engage contemporary physicists, he situates their findings (if at all) as a possible 'crisis in physics'.[40] It presents a challenge as it entails 'a crisis for a future physics, in other words, none other than a crisis for a future ideology'.[41] That he conflates physics and ideology in this way appears as a rather anti-intellectual deflection of science of physics so as to assert a Marxist historical materialism, which as a science must be seen today as incomparable to physics. Yet, we must remember that in Tosaka's day, he would have been surrounded by nationalists and this Marxist ideology provided him with a means of maintaining critical cognizance. Nevertheless, he would eventually recognize the impact of quantum physics in a somewhat more noteworthy way.

In the following passage, note how Tosaka refocuses the Kyoto School (Nishida and Tanabe) attraction to *contradictions* within a Marxian frameworks as historical conflict that would for him lead to revolutionary change.

Translation

from 'The Natural Sciences and Ideology: What Is the Significance of the Crisis in Modern Physics?' (*Shizen kagaku to ideorogī —— gendai butsurigaku no kiki wa nani wo imi suru ka* 自然科学 とイデオロギー── 現代物理学の危機は何を意味するか, 1934)[42]

Though I would not go so far as to borrow Aristotle's classical expression,[43] all things change, move and develop toward *contrary* [*hantai*] things. Thus, indubitably changes in a thing are manifest in the phenomenon of drifting toward its opposite, because the dynamics of this change, the dynamics of movement, exist contradictorily.[44] Even seen in terms of this perspective, one can claim that change in all things is based on contradictions. The motives for people to try to refuse these new concepts of time and space derive none other than from a realization of there being little relief from the contradiction, simply put, found in the opposition of the old and new approaches to these concepts. Yet, if they just resulted in contradictory opposites, movement [to the new concepts] would never arise therein, only a suspension of movement. Therefore, it is to be expected that people have not acknowledged a movement from the old concepts of time and space to the new ones. In order for contradiction to constitute the dynamics of movement, contradiction must not stop in contradiction but on the contrary proceed to resolve the contradiction, as the essence of contradiction always lies in its necessary sublation. In this way, two things that form a contradictory opposition commence sublation, are unified as two items in a transition – a movement – of both the former and the latter so that they can be again distinguished. Now, upon abstracting this result only from this process, it appears as the phenomenon of *contraries*. What resolves a contradiction with its opposites is a *unity of contradiction and opposites*. It is dialectical progress. The concepts of time and space progress by way of relativity theory into a typical *dialectical development*. The crisis of time and space must be characterized in part as its development. However, this crisis has already been overcome. That is, the new – relativistic – concept of space-time as a *specific* synthesis of itself, takes the old – absolute – concepts into a synthesis considered in light of each,[45] and, in short, as one moment of itself due to ongoing sublation. It is not that space and time have unravelled, but that these concepts have *progressed*.

Therefore, the concept of relative space (and time) is certainly not mechanically and immediately in conflict with the concept of absolute space (and time) as usually imagined. Relativity in the space (and time) in this case indicates a unity of the absolute and relative. To such relativity we generally confer the name *dialectics*. Consequently, space (as well as time) now is understood as something *dialectical* because of relativity theory.

It goes without saying that we must similarly pay attention to this feature in the relation between space and time. Actually, according to relativity theory, as mentioned above, two things cannot be understood independently, but can only be grasped as a unity of both of them. Moreover, a unified time and space here obviously does not permit any sort of merging or unification, as time and space remain incontrovertibly opposed to each other – with the *relativity* between them forming a *dialectic* between them.

The dialectical grasp of space and time must be resolved in a dialectics of *movement* making a concept that unifies them. The theory of relativity thus seems fully given over to this demand. The theory of relativity historically emerged out of a negation of absolute movement.[46] The demand is for the necessary unification of the movement of A in relation to B and the movement of B in relation to A. The movements of neither A nor B are privileged relative to one another. Both are relative. But, that is not all. It also becomes meaningless to speak of discrete independent movements with respect to the movements of A or B, as in a sense movement is nothing but the movement between A and B. The movement of A and the movement of B as the movement *of A and B* is unified. The depiction of this unity constitutes a coordinate conversion system – according to a conversion of space (and time) coordinates – a system by which the movements of A can accommodate the movements of B.

* * *

Following relativity theory, physics has been pushed out of the non-dialectical – what we call the *metaphysical* – into dialectical stages. Wondering about what category it falls under in being made aware of facts is not the problem here. In considering the fundamental concepts in physics – including spacetime and matter – as if they confront a crisis on this occasion, was nothing but an expression of the conservative pangs of having to convert metaphysical thought into dialectical thought.[47] According to dialectical thought, this crisis has already been overcome. Regardless of whether one is aware of it or not, this progress in physics can likely only be understood as progress to the degree we rely on it [dialectical thought].

Nevertheless, while we arrive at this conclusion, it is not considered to have been endorsed with any favour by accomplished physicists or philosophers. The number one reason they oppose it is probably the following. More than anything, what relativity theory has achieved is the mathematization and geometrization of physics. Because of it, physics has been able to become all the more methodical, theoretical and exacting. Now – given their assertions – physics and its incorporations of mathematical and geometrical methods are absolutely not dialectical in methodology. So far as it is not so, is it not in itself something dependent on a most content-based [*naiyō-aru*] *formal logic*? Then, might one wonder whether formal logic might be the diametric opposite of dialectics?[48] But, our answer is simple and clear. While dialectical thought – dialectical logic – is opposed to formal logic, far from absolutely rejecting it without mediation, it sublates formal logic as part of its moment. Still, it amounts to no more than the rejection of formal logic's unconditional adoption. Therefore, according to relativity theory, physics's far-reaching adoption of mathematical – formal logical – methods is not the same as excluding dialectics per se. On the contrary, actually this comprehensive adoption of mathematics (geometry) in fact results in physics becoming completely dialectical. Adoption alone of the formal logic of mathematics itself renders such a dialectics concrete.[49] If there are still people who do not agree with this, those kinds of people, instead of seeing dynamic fundamental factors in the progress of scientific methods and research, see science as nothing but some fixed abstract schematic – just see the above.

* * *

In light of the new physics, what is considered the disappearance [*shōmetsu*] of matter is actually not its *disappearance*, as here once again one must attain an understanding matter *dialectically*. That matter has been reduced to waves while also being able to be grasped as independent of light[50] is not due to matter being an isolated fixed object, but rather actual proof that matter – for instance light – exists only in relation with its object. To say that matter is a probability is to declare that existence – which is matter – is something simply – mechanistically – *deterministic* (see below). The new developments of the original quantum theory as a whole themselves, in terms of quanta of existence, as I expect we already have seen, are motivated out of a dialectical unification of continuity and discontinuity.

Yet, while idealism uses the fruits of the new physics, the data [*zairyō*] furnished in materialism – that is, Marxism – does not include only problems of matter. Though we arrive at matter today by way of space-time and motion as fundamental concepts in physics, we now face problems concerning the *principle of causality*.

* * *

In terms of so-called determinism, we must be aware that causality, necessity and contingency all derive from *mechanistic determinism*.

* * *

Given that so-called determinism assumes existence to be a mechanistic assemblage [*shūgō*] of various situations, on top of a card on which such an assemblage is imagined and so established, necessity and contingency become opposed in such a way that neither of them can move. In this case, to say contingent things become necessary can in principle only be said within a [given] scope for the cognition of existence [*sonzai no ninshiki no shūi*], since existence itself has nothing to do with anything.[51] Necessity and contingency are mechanistically – metaphysically – opposed. Today this sort of mechanism runs counter to a fundamental result of the new physics – the uncertainty principle. Therefore, the new physics compels people to give up mechanistic determinism and thus also mechanistic necessity and contingency. In brief, it demands the abandoning of mechanism. Thus we must by all means adopt *dialectics* in opposition to mechanism and as a substitute for it, adopt *dialectical determinism* as a substitute for mechanical determinism, and adopt *dialectical necessity and contingency* as a substitute for mechanistic necessity and contingency. The pressure from the facts of the new physics demands it.

Necessity and contingency as understood dialectically are indeed genuine necessity and contingency. Moreover, only this sort of approach to necessity and contingency can generate a substantial unity of relations [*jisshitsuteki ni renkan-tōitsu*].[52] Then, what is the unification of necessity and contingency? It is the *law of cause and effect understood dialectically*.

Both necessity and contingency can at present only make sense if we assume the causal. The causal is simply the necessary and that which is not contingent. The causal must take the form of a unity of the opposition of causal necessity and causal contingency. Then it goes without saying that what is clear here is that what has broken down in accordance with the uncertainty principle in the new physics is not the *law of cause and effect* itself but rather nothing but the *concept of mechanistic causality*. On the contrary, a genuine law of cause and effect – a law of causality properly understood according to a concept of dialectical causality – must first be capable of being put forth clearly.[53]

* * *

Various fundamental concepts applied in physics – and the natural sciences in general – are in the end based on philosophical categories, which are themselves ideological, of a worldview, theoretical and, thus, also logical formations [*soshiki*]. These philosophical categories of course are not limited to specific uses in the sciences, since they are of general use. Similarly grounded basic concepts in the natural sciences are also to this extant not necessarily limited only to particular things with special uses in the natural sciences. So, for example, basic concepts in economics directly correspond with those in physics and can have ideological intermingling. In such cases, the deep ideological – political – nature of economics can even be reflected just as it is as the ideological nature of physics.[54] Theoretical physics in this way at times is able actually to be *reactionary* [*handōteki*].[55] Actually we have already touched upon several examples along these lines.

Thus, interior moments in the natural sciences[56] – its theoretical and methodological structure and basic concepts – and its exterior moments – the historical and social existence of the natural sciences – can begin to be synthesized into unities while serving as mediators of ideological existence.[57] What is called a crisis in modern physics – in the natural sciences in general – is in fact nothing but a [mechanist, metaphysical] crisis of bourgeois ideology. Thus, to have to take note of the collapse of this – full of flaws to begin with – ideology as some sort of *crisis* is thus to be aware of the process of the collapse of capitalism as a crisis of society itself. To the extent that one is aware that the process of the collapse of capitalism is the progress of society itself, there is no sort of crisis in modern physics, just some rapid progress in it. This is the subject of the crisis in modern physics. Over the last few years, capitalism's deadlock has become more obvious, so that the particular announcement of a crisis in physics are thus certainly not coincidental.[58]

> **Note.** We have to allow for the general inclusion of Rudas's standard approach without editing it out of our analysis. Although Rudas was a leading member of Deborin's faction, his criticism from within the Soviet alliance is of no direct relevance in this case.[59] [Tosaka's addendum.]

Questions for further discussion

Note how the style of this last paragraph, immediately above, seems superficially to emulate Nishida's spiralling prose that accrues through careful situating of concepts and linking of each conceptual building block coherently. By contrast, Tosaka is better described as taking glancing approaches to the façade of a situation and nearly always pulling back to a safe distance before delving too deeply, so that he offers an ideology, a Marxist worldview that contains categories by which to situate all things encountered in the world according to *a priori* givens.

1. Both Tanabe and Tosaka explore a 'crisis in physics'. How do their approaches differ?
2. Drawing on Tosaka's positions on 'nature dialectics', how would he likely respond to Tanabe's uses of dialectics?
3. Does Tosaka's presupposition of a Marxian ontological and epistemological horizon limit his credibility as a philosopher?
4. Does not a philosopher's credibility and force of argument in part depend upon carefully establishing, from the ground up, any *a priori* assumptions?
5. Taking the definition of key terms as an example of such work, how do some words get repeated to the point that all roads lead to the dialectical?
6. How do Nishida, Tanabe and Tosaka each distinctly situate dialectics in relation to modern physics?
7. Ultimately, what do we learn from Tosaka's adherence to questions of political philosophy to address questions in the philosophy of science?
8. Tanabe writes:

> As Dirac himself has been saying all along, the scope for a relativistic development of quantum mechanics has yet to be established. We cannot even begin to predict future developments. However, just in terms of what we now know to be the structure of physical existence, we cannot conceal its dialectical foundations. To this extent, the so-called nature dialectic too must be said to hold a certain currency that should not be dismissed.

Here one might ask if Tanabe is referring to the same 'nature dialectic' Tosaka championed in his writings and in support of others in his *Yuibutsuron kenkyū* (Studies in Materialism) journal can we distinguish their two approaches? Defend one approach over the other.

References

(Tosaka's internal references in notes are not listed.)

Doak, Kevin M., 'Under the Banner of the New Science: History, Science, and the Problem of Particularity in Early Twentieth-Century Japan', *Philosophy East and West* 48, no. 2 (April 1998): 232–56.

Engels, Frederick, 'Philosophy of Nature: Cosmogony, Physics, Chemistry' (1877), in *Anti-Dühring*, https://www.marxists.org/archive/marx/works/1877/anti-duhring/ch04.htm.

Kant, Immanuel, Paul Guyer and Allen W. Wood, *Critique of Pure Reason*, New York: Cambridge University Press, 1998.

Kōsaka, Masaaki, *Japanese Thought in the Meiji Era*, Tokyo: Pan-Pacific Press, 1958.

Nishida, Kitarō (西田 幾多郎), *Nishida Kitarō zenshū* (Complete Works of Nishida Kitarō), 19 volumes, Tokyo: Iwanami, 1978–9.

Tanabe, Hajime (田辺 元), *Tanabe Hajime zenshū* (Complete Works of Tanabe Hajime), Tokyo: Chikuma Shobō, 1963–4.
Tosaka, Jun (戸坂 潤), *Tosaka Jun zenshū* (The complete works of Tosaka Jun), 5 volumes, Tokyo: Keisōshobō (勁草書房), 1966–7.
Tosaka, Jun et al., *Tosaka Jun: A Critical Reader*, Ithaca, NY: East Asia Program, Cornell University, 2013.
Watsuji Tetsurō, 'The Significance of Ethics as the Study of Man', trans. David A. Dilworth, *Monumenta Nipponica* 26, no. 3/4 (1971): 395–413.

5

What We Can Learn from the Kyoto School

5.1 The philosophy of physics and competing conceptions of materiality in Nishida, Tanabe and Tosaka

New approaches to matter and materialism in the last two decades attempt to facilitate the withdrawal of anthropocentric biases from how we recognize materiality. With the advent of modern physics, philosophy and science entered into a rethinking of the ontological (questions of being and becoming), the phenomenal (questions of experience and intentionality) and the epistemological (questions about knowledge formation), forcing the rethinking of materiality today in light of a bigger picture and longer view of debates in both; however, such work has yet to be explored in dialogue with Japanese philosophers, who have their own competing histories of situating substance in light of competing schools associated with various Indian, Chinese and Western philosophies in Kyoto School writings and beyond. A genealogy of material substance that includes a more global range of approaches to philosophy and science now seems possible as we translate and read these texts and debates already in dialogue with contemporary philosophers and scientists who contributed to interdisciplinary discourses in the 1910–1940s period.

Dialectics in the Kyoto School

As we saw in Figure 2.1, Nishida pursues a quasi-dialectical relation of subject and object that he painstakingly modelled in what he called a 'schematic explanation' of his system of thought. Nishida's earlier work (presented in 2.1) situates the subject objectively, in light of Fichte's fact-act (*Tathandlung*) and absolute ego, focusing on *poiesis* (active co-emergence with the world, applying for Nishida to scientific experimentation as well as artistic creation) to unify his system.[1] Fact-act *poiesis* animates subject and object within a relationality that encompasses active engagement of sense and bodily faculties as well as entering a common dimension of objects *as* an object oneself. Thus, one might ask, how did Nishida's situate relationships of *relationality without a priori being* as neither a categorical substratum (to be compared with Kant, Tosaka and Tanabe) nor a physical given. Inspired by Buddhist thought but developed mostly in conversation with European philosophy as known at the time, Nishida began with emergence in the present and predicated on selflessness (*muga*) in a void (*kū*), while developing his system in rational, scientific and philosophical terms. Concomitant with the downplaying of the self, Nishida would then soon develop the concept of the site or *basho* that becomes a universal locus insofar as it is dialectical. Nishida would then situate the self as both human self (itself) and world itself (itself), and establish various dialectical relationships that are characterized by 'mutually contradictory self-identity'. The *basho* becomes but one important nexus in Nishida Philosophy for illustrating, developing and indeed dramatizing such relationality. At stake are relations potentially anywhere there are acts and worlds.

Retaining aspects of Nishida's defining self-less void-based dialectics situated in matrices of place, Tosaka may be seen to focus on the transition from *a priori* systematic metaphysics (Kant) to a multiplicity of recognizable spaces and spatialities that challenge Nishida's dialectics from a Marxian perspective. Marxism for Tosaka is not simply an inverted Hegelianism (as Marxism is often called) but a form of critical demythologization in his social criticism focused on questions of ideology and common sense in terms of social function (praxis, practices) in the creation of classes and so forth. We might ask how (or even whether) Tosaka recasts Kyoto School premises regarding the status of *a priori* givenness of the world by rendering all within Marxist polemics and refusing to allow the social to bow to science. He writes that '*cosmic time* provides an *order* of all of existence (including not only nature but also the history of human society). This corresponds to the so-called spine of a nature dialectic ... the *fundamental laws* of nature have followed a nature dialectic in terms of the movements of the changes, developments, and transformations of nature'.[2] In such passages Tosaka displays how indebted he is to his early studies of space that still remain comprehensible primarily in classical Newtonian terms. As such, written in the 1930s, they thus demonstrate Tosaka's refusal to situate dialectics in a post-causal way, given that quantum mechanics relied entirely on probabilities by this time to situate matter. His voice is outspoken, but to this very degree it also relies on a form of certainty itself problematized by the new physics.

One way of framing the diverse strands of Kyoto School writings on modern physics is to consider how they emplace the observing subject vis-à-vis dialectics, a common thread interpreted distinctly by each. How does each philosopher situate dialectical relationships in the very conceptualization of relations between subject and object in light of the new physics? Over time, what changes can one see in their relation to science and the philosophy of science within their writings? If one frames each of the three philosophers around such questions, an overview becomes possible and a broader framework for analysis makes itself apparent.

World philosophies, null ontologies and modern physics

Can we consider Nishida and Tanabe's exploration of null origins of local ontologies a sort of missing link for bringing world philosophies into conversation at the common nexus of modern physics? After all, Nishida and

Tanabe have dedicated their thoughts and schematic forays into definitions of modern modes of relationality in ways that stand apart from the work of others attempting similar definitions in ontologies and epistemologies inflected by modern physics. Yet unsolved problems resulting from quantum physical demonstrations pioneered by Bohr, Heisenberg, Schrödinger and Dirac still exist, and the debates of their era are still relevant. Nishida and Tanabe situate relativistic and quantum approaches to the particle/wave nature of matter as exhibiting beginnings always already within a framing void that places the burden of emergence on what is introduced minimally, both in the sense of the development of a null-dimensional substrate within which the human subject moves in Minkowskian world-lines confined to the locality relativity demands, in Nishida, and in the sense of the refinements of Dirac's holes (and discovery of negative protons and antimatter), in Tanabe.

Nothingness and frames of ontological emergence inflected by modern physics: contemporary questions for Nishida, Tanabe, and Tosaka

One may briefly examine treatments of nothingness-oriented dialectics in Kyoto School philosophers Nishida Kitarō and Tanabe Hajime, who engage questions of space from Hegel to the implications of quantum mechanics. Nishida and Tanabe each offer approaches to space and matter for the new physics in ways that dovetail with our contemporary new materialism's concern that also engage the implications of quantum theory in relation to space and matter. For instance, when Diana Coole and Samantha Frost (proponents of new approaches to materialism in socio-political thought) entertain 'choreographies of becoming that ... find cosmic forces assembling and disintegrating ... objects ... within relational fields, bodies composing their natural environment in ways that are corporeally meaningful for them',[3] they frame *some scales* in the formation of these *objects*. But, what if we recognize one of the implications of quantum physics, namely that elementary particles may be *anywhere* and that even atoms are not essential to matter so much as illusory simplifications of complex quantum relations ascertained only probabilistically? Then, for quarks blinking in and out of space or for stars forming in the farthest reach of time that our telescopes can now access, scale and relation of parts are both problematic. *How* particle-waves blink in/out is thought to be bound up in the earliest parting of the cosmos (big bangs) and subsequent star formations and supernovae; matter everywhere *is* bound up with – entangled with – matter potentially anywhere (including galaxies far away).[4] The matter of disentangling objects at the subatomic level – merely to know them – may be difficult, based on probabilities, not absolutes, as entanglement and non-locality of the wave-particles of matter are universal and sites of investigation are very much besides themselves and subject to changes in the observation process itself. The role of the recognition of awareness of frames of structurality (poststructuralism) and how various agencies weigh in on their formation is not necessarily in opposition with issues of framing materiality.

Quantum physics does not itself offer the last word on how we exist dialectically, but it does complicate and undermine how we conceive space. We have a means of conceptualizing how we are no longer 'lost in space' but rather have always been, in Wittgenstein's sense, 'held captive to a picture' of space and Cartesian coordinates with the human vantage holding sway. As Kristian Camilleri, a leading Bohr specialist, argues, 'the concept of the electron's location in space ceases to have applicability in quantum mechanics. In this way, the very notion of an "object" in space and time is recognized to be an idealization with limited applicability, but one which, for Heisenberg, ultimately remains indispensable for the possibility of human knowledge.'[5] Heisenberg recasts Kantian *a priori* transcendental concepts from a metaphysical structure into a 'practical one'.[6] *Space* thus becomes a practical concept with a 'limited range of applicability' rather than an absolute given. For Niels Bohr, space must be recognized as destabilized, subject to radical reduction, since 'the representation of a state of a system can never imply the accurate determination of both members of a pair of conjugate variables q and p ... there will always be a reciprocal relation'.[7] The position and momentum of an electron can only be known 'by referring to the mutually exclusive conditions for the unambiguous use of space-time coordination, on the one hand, and dynamical conservation laws, on the other'.[8] Bohr emphasizes how this is the source of complementarity in quantum physics, but implies that quantum relationalities still require the scientific culture that has provided us with a manifest world based on classical physics. This process of renegotiation limits our 'freedom of constructing and handling the measuring apparatus, which in turn means the freedom to choose between the different complementary types of phenomena we wish to study'. This, he implies, blocks our access to a clearer 'quantum-mechanical mode of description', one that might situate itself in terms beyond the limited purview of classical physics. The basic scheme and formulae used in testing quantum events are 'mathematically consistent'[9] while yet to be described in a language more native to quantum physics.[10]

Since the work of physicist John Bell in quantum theory in the 1960s, the emphasis on the destabilization of events in continuous space-time has been losing ground to propositions that the subject–object distinction *itself* has inhibited understanding of quantum observations. Bell argues for local be-ables (or maybe-ables), written 'beables', to account for the limited capacity to engage local events without having to re-establish a master narrative of space-time relations in the wake of the 'New Quantum Theory', dominated as it has been in the philosophy of science by the Copenhagen Interpretation. Empirical data has led to the relegation of space-time to that of an incomplete and unreliable concept. What remains *be*able is local engagement. 'Beables' are in quantum theory among the more plausible modest ontological postulates in light of where physics has placed us today. As Newtonian science helped Kant, Hegel and even Nishida see the world in relation to force, quantum mechanics suggests models not bound to metaphors of force, which notably proliferated in the age of colonialism (which our thinking has yet to be extricated from). Beables, as hybrid quantum-material places, suggest the need for new models of the dialectic of being and nothing that places the onus on substance to establish itself after the fact of nothing – a reversal of a long-held assumption in Western philosophy. Thus we live in an age of a dialectics of empty space. As Bell argues, a '"system" under study' may simply be defined, rather, 'as a limited space-time region'.[11]

Maintaining that 'local *beables*' offer possible ontologies produced at sites of attention, Bell thus refuses to police the subject–object distinction or to contain it in new procedures. This orientation toward remnants of space as the product of studies in physics avoids (critics would say ignores) being sidetracked by many related issues in quantum theory. Nishida's own site-specific dialectic provided a frame for direct engagement with quantum physics by his successor, Tanabe. Could their work offer various means of situating a dialectics of emergence in a sense commensurate with Bell's interpretation of quantum theory, a minimal starting point working out of a Buddhist ontology expressed in world-philosophical intertexts and dialogues? Both Nishida and Tanabe suggest alternative conceptions of coexistence – historically, socially and physically – in interacting but isolated individual points of access not forming an objective (or intersubjective) whole. The idea of superpositioning, the simultaneous existence of observables associated with a given (repeatable) phenomenon that nevertheless cannot be reduced to one phenomenal measurement due to the role of the observing means and apparatus in determining them, presents a duality that is fundamental to quantum mechanics and is situated differently by each of these three Kyoto School thinkers.

Nishida Philosophy and Tanabe's writings on quantum mechanics will now form the basis for an exploration of dialectics and space in light of contemporary discussions of quantum theory within philosophy. Tanabe suggests that Nishida's dialectics of nothingness, one of the key concepts inspiring Kyoto School philosophy, was supported by the new quantum theory. Nishida's idea of 'absolute nothingness' (*zettai mu*) becomes the basis for situating seemingly *any* dialectical opposition as a fundamental contradiction that structures ontologies differentially. Nishida writes, 'When the absolute opposites form a self-identity as "one in many" and the "many in one", an infinite amount of particulars are determined mutually vis-à-vis each other.'[12] Kyoto School methods variously begin with an undifferentiated void to produce a processual, productive dialectic not subject to synthetic being, but situated in terms of contradictions. As James Heisig argues, precisely *because* of the focus on nothingness as an empty set problematizing anything that enters an ontological field of potentiality and epistemological void, it undermines essentialist identity of matter and objects. For Nishida, 'To call reality itself *absolute nothingness*' is to say that 'it is subject to the dialectic of being and not-being, that the identity of each thing is bound to an absolute contradictoriness'.[13]

Gereon Kopf sees non-dualistic, non-essentializing difference as 'the driving source behind [Nishida's] seemingly infinite dialectic'.[14] It is this development that forms a supplement to our understanding of superpositioning (as expressed in complementarity and the uncertainty principle) and the problem of describing space, since Nishida 'maintains that the multiplicity of phenomena cannot be reduced to a oneness, be it a self-identical being à la Spinoza's substance or a self-identical place à la the *basho*'.[15] Nishida grounds the apparent void, that quantum mechanics brings to light, at a site predicated on nothing while subject to ethical concerns. He also offers a use of 'discontinuity' (*hirenzoku*) or 'continuity of discontinuity' (*hirenzoku no renzoku*) 'to undermine a causal-mechanistic approach that denied the possibility of free will and creativity, and to accentuate the dimension of the world of engagement'.[16] Following this logic, decohering emplacement in space (given our understanding of quantum complementarity) might obtain a model for situating freedom to the degree it carries the theoretical potential to undermine predicted causes and expected results – so crucial to biopolitical control today, where predictability governs capital markets.

One may also situate these Kyoto School dialectics in the philosophy of science in relation to Hegelian dialectics. Hegel writes in the *Logic of Science* that 'Existence proceeds from becoming. It is the simple oneness of being and nothing ... It is at first in the one-sided determination of *being*; the other determination which it contains, *nothing*, will likewise come up in it, in contrast to the first.'[17] Without arguing *why* being is determined first, he simply reflects a long-standing consensus in Western philosophy since Plato that being, not nothingness, must be the foundation of any ontology. It is this suppressed nothingness that forms Nishida's basic point of departure in redefining Hegel's dialectical thinking. His use of nothingness, however, was inspired by the use of emptiness in Indian and East Asian Buddhist thought. While *emptiness* per se is rarely discussed in Nishida (until his very late period), it is a feature of his foundational philosophical method present in his use of 'nothingness', which is inspired by and often considered a formal philosophical articulation of the Mahayana Buddhist concept of *emptiness* (空). 'Emptiness' is a key concept here, pointing to an ideal *openness* to recognizing that the world is not literally materially empty; rather, it is apprehended with the possibility of *seeing through* the material obstructions that form conflicts and force-relations in the world. As such, he turned to redefining a systematic dialectics akin to Hegel but inflected by the Buddhist concept of emptiness. Although treatments of Nishida in Western languages often highlight the Buddhist legacy in religious terms, it is important to underscore that Nishida wrote philosophy of the most abstract and logically ordered variety. Thus 'empty' becomes a metaphor for the obligation to account for spaces (scales, scopes, object-situations and relations) without resorting to modernist paradigms of instrumental reason in conjuring ontological emergence. Here we also see how situating the philosophy of science complicates and complements how we map out world philosophies too, as they are in processes of becoming that depend on the frameworks we use. Lucy Schultz builds upon James Heisig's interpretation of absolute nothingness in Nishida, arguing:

> Nishida sought to articulate a rational standpoint that could account for itself without being contingent upon conditions determining it from the outside. Only absolute nothingness can satisfy this criterion. And yet ... [i]t has no existence in-itself distinct from the myriad manifestations of the historical world. The place of nothingness is no place other than the medium of concrete reality.[18]

The impulse for Nishida's work was in part based on an attempt to supplant a metaphysics of being in Western philosophy with one of nothingness found in Buddhism and avoiding nihilism. Schultz focuses on how Nishida's affinities with Hegel reflect a common orientation toward the concrete in variations of a 'dialectical ontology in which being itself is shaped by the dynamic interplay of the self and world'.[19] Nishida's dialectics thus withdraws from Hegel's historical idealism by reworking various dialectical oppositions as a *present* that refers to nothingness as its base premise rather than being. Nishida maintains a focus on structuring contradictions that form the emptied non-present nevertheless still substantive – albeit tentatively balanced and changing – as a dialectical present subject to self-contradictory productive forces that preserve a role for free will in the framing of the frame.

While Hegel repeatedly refers to space and time as being of the same cloth, what concerns us, with respect to the task of understanding space after quantum theory and how matter is present, is Hegel's own use of *place* to situate space and time at a specific place, which is so essential to Nishida. Hegel argues in a way that seems to sheer away any sense of Kantian *a priori* transcendental categorical scaffolding:

> Space in itself is the contradiction of indifferent juxtaposition and of continuity devoid of difference; it is the pure negativity of itself, and the **initial** *transition into time*. Time is similar, for as its opposed moments, held together in unity, immediately sublate themselves, it constitutes an immediate *collapse* into undifferentiation, into the undifferentiated extrinsicality of *space*. **Consequently, the negative determination here, which is the exclusive point, is no longer merely implicit in its conformity to the Notion, but is posited, and is in itself concrete on account of the total negativity of time. This concrete point is place.**[20]

This passage appears crucial to our understanding of Nishida's use of place (*basho*) as a major concept in the development of his later thought as well as reinterpretations of Hegel with implications for critical materialism today.

The key elements of Nishida's place are here: the focus on space as a repression of time's other places, with concrete focus on *place* enabling an escape from abstraction and a point of access to being and nothingness. While Nishida would clarify the 'contradiction of indifferent juxtaposition continuity devoid of difference' as a *working model* for his system, replacing the movement toward substance with a movement toward nothingness *in order to highlight contradictions as unity through interaction, rather than the hierarchical overcoming through sublation* (and progress of the human spirit of overcoming). He writes the following in one of his last essays on the logic of the place, invoking Leibniz and Kant within his own system:

> I am an expressive monad of the world. I transform the world into my own subjectivity. The world that, in its objectivity, opposes me is transformed and grasped symbolically in the forms of my own subjectivity. But this transactional logic of contradictory identity signifies as well that it is the world that is expressing itself in me. The world creates its own space-time character by taking each monadic act of consciousness as a unique position in the calculus of its own existential transformation.[21]

This consciousness is realized through its space-time character not based on essence but rather in the present itself in relation to others, but self-determined, and limited by the foundational assumption of nothing (rather than being) as well as the role of conscious choice of frame: 'As humans, through the contradictory self-identical character of time and space, we transcend the determined, causal world of the here and now as self-determinations of the absolute present itself.'[22]

In his essay 'Place' (*Basho*) Nishida situates space thus: 'When one considers space-time, and force all as means of thinking, the objective place [*kyakkanteki basho*] in which the experiencer oneself is situated would seem to be akin to a field of transcendental consciousness.'[23] Nishida is careful to insert qualifiers to underscore how his 'logic of place' is a critical metaphorical apparatus in its own right for overcoming the dualisms which limit Western philosophy heretofore to binary oppositional dialectics (Hegel) or *a priori* transcendental categories (Kant), while accounting for free will and mechanisms of determination that are not reducible to raw forces. As mentioned earlier, Nishida's interest in recuperating a subtly understood free will and creative *poiesis* is complemented by complication of the subject and object in modern physics (in both relativity theory and quantum mechanics). Nishida shares an interest with our contemporary new or critical materialist focus on discerning means of conceptualizing knowledge without privileging anthropocentric cultural perspectives that may include the production of certain types of space themselves. He argues, alluding to Kant, that just as we cannot posit something resembling a thing, we cannot posit something resembling force because in both examples they 'are objectified by discerning subjects'.[24] Nishida situates such contradictions attendant to both force and thought at a site (*basho*) in relation to nothingness according to the idea that, 'When a place is truly nothing, such contradictions vanish, and we again see individual independent existences as if things in space.'[25] As Krummel argues in his discussion of this essay, '*basho*' (place), being named after Plato's *chōra* (place or locus), comes to function as a dynamic matrix of forces within various dialectical oppositions. Nishida 'transposes the Platonic ideas into epistemological categories that form sense-matter, and chōra becomes the place *qua* field of consciousness … for that interrelationship of form and matter'.[26] Here *nothing* operates as a bottomless receding framework accommodating a spaceless spatiality for potentiality and thought. As such it seems a neutral placeholder for determinations (judgement) and a locus for volitional action interrelated, restrained and defined by others (human and non-human).

Nishida himself presents a limited engagement with quantum theory and yet may be understood as conforming more closely to classical physics (based on force, mass, inertia), following Bridgman; however, Tanabe does engage questions of dialectics and the new quantum theory by extending Nishida's approach to dialectics.[27] Tanabe recognized how determinism, simple causality so fundamental to classical physics, lost its ground with quantum physics. Even though many physicists would assert Schrödinger's formula to be a good estimator of a particle-wave's state and potential for movement, the formidable uncertainty principle (the Copenhagen Interpretation) undid the basic laws of causal relation in space, now reduced to relations of probability that engender infinite possibilities and in effect naturalizes the ubiquity of possible superpositions of complementary states to a situation.[28]

Recognizing that quantum physics had supplanted relativity theory as the driving innovation in the field of physics, Tanabe wrote, '[i]n both its form and contents, the objectivity of physics has a dialectical structure' based on its development' and, immediately in the next chapter, argues that these *new physics* themselves are in a dialectical relationship with classical (Newtonian) physics.[29] Thus 'nature in physics is as a whole universally subject to a dialectical structure'.[30] Tanabe sees the impulse to recuperate classical physics as displaying an unwillingness to embrace the dialectical challenge that quantum theory itself poses, what should be recognized as a new direction in physics. Thus he argues that 'the revolutionary significance has yet to be adequately accounted for'.[31] He also realized the either/or option of quantum versus relativity theories was a red herring, writing that relativity theory only rethinks the relation of geometric (Euclidean) space and classical (Newtonian) physics, but does not extend its investigation to 'the structure of physical existence itself'.[32] Tanabe realized that the shift from classical to quantum physics could be traced in the shift from treatment of mass, space and inertia to questions of measures of frequency and wavelength.[33]

Many approaches to quantum physics have arisen since the foundational debates focused on Einstein and Bohr, yet scant attention has been paid to Tanabe's work. In 'Between Philosophy and Science', Tanabe cleverly maps the 'micro-probabilistic theory [of quantum mechanics] as the basis for a dialectical ontology, which must be understood as something realized through experimental facts'.[34] This dialectic is his response to reading Dirac and others on the undecidability of wave/particle delineation of matter and its implications for the basic understanding of how existence presents itself as well as spatial coordination on the level of predicting energy levels of positively/negatively charged electrons in relation to void-holes, how photons exhibit superpositioning, and related issues.[35] Statistical probability becomes the most viable means for resolving these ambiguities borne of the loss of classical and relativistic models of space after quantum mechanics. While mediation based on both *theory* and *actuality* may form a dynamic unity, in Tanabe's essay 'The Development of Operationalism in Quantum Theory' (*Ryōshiron ni okeru sōsashugi no hatten*), he writes that 'there is no one existence that would mediate the two beyond it. Rather, we should call what mediates them absolute nothing [*zettai mu*], a principle of interchanging negative unity.'[36] Thus Tanabe falls back on a central concept of post-Hegelian nothingness (along the lines of Nishida) as a premise for an access point in the present in order to ground his interpretation of quantum physics. By way of quantum theory's capacity to 'make physics and philosophy self-aware of mutually exclusive complementarity [*sōhaiteki sōhoteki*]'[37] it allows us to see the precarious and by no means dominant (albeit necessary) role of non-dualistic mediation of partial, probabilistic spaces predicated on nothing rather than being. However, further elaboration of Tanabe's own thought in relation to Nishida's in light of modern physics would entail a much longer study.

Questions for further discussion

1. In light of the above discussion of place and nothingness in Kyoto School versions of the emergence of self and world, how can one reimagine the concept of the Anthropocene (the current geological age defined by human dominance of the planet)?
2. How might Nishida's schematics be used to situate beables?
3. Can Nishida's emplacement-orientation offer a more dialectical accounting for the quantum mechanical (probabilistic) determinations and limits to understanding our world in light of the complementarity of the uncertainty principle?

5.2 Implications for critical materialisms and post-humanism today

Situating physical systems: competing forms of quantum intuition

In its matrices and formulae for comprehending observable correlations of data and states of matter, practical applications of quantum physics have thrived for decades without apparent need for resolving classic and quantum worldviews. However, these means of determination always involve the incorporation of unknown (hidden) variables in their matrices. This absence of coordinate axes that conform to ordinary space makes the production of space itself problematic. Rather, what is produced in quantum work is confined to abstract vector spaces (two-dimensional Hilbert space) that may or may not coincide with the manifest reality of everyday coordinate and perspective space that we move through as urban entities.

To circumnavigate the classical–quantum divide in physics, Nishida and Tanabe may be situated in the philosophy of science as describing frames that offer fresh ethical grounds and genealogies based on non-binary mediation in establishing the ground for the recognition of non-binary mediation by way of elements of modern physics. They offer an alternative road to critical materialist agential realism, which in adhering to Western assumptions of being (relative to the Kyoto School adherence to a void-substratum) tends to exacerbate the binary extremes supposed to be dissolved in a realism based on splits and divisions rather than multi-participant actants in dialectics of emergence out of a void as such. Agential realism sees 'entangled ontologies' rather than a classical 'mirror image' 'sameness, mimesis' and 'separate entities'.[38]

A recent thought experiment published in *Nature Communications* clarifies the long-standing problem of having concurrently recognized possible frames for comprehending quantum system observation, in effect drawing quantum theory closer to the sort of grand dialectics Nishida and especially Tanabe suggest for the new physics. The paper sheds light on the problem of complementarity in speculative extrapolations and applications outside modern physics proper and concludes that 'quantum theory cannot be extrapolated to complex systems, at least not in a straightforward manner'.[39] It suggests that quantum physical scales at the subatomic may not be helpful in mapping macrological objects such as biological cells or even complex molecules. This article reflects a growing recognition of dynamic complexity in the range of interpretations of such competing frames[40] and, moreover, compels us in our interdisciplinary uses in the humanities to update how we relate necessarily metaphorical language and formulae in staging our human understanding of observations of quantum systems and the roles of various technology, including quantum computers themselves, which nevertheless should be clearly understood to be tools that have no role in themselves in the central decision-making concerning quantum scales of applicability. As Frauchiger and Renner put it, Schrödinger's acquiescence to a view that 'macroscopic superposition states do not represent anything contradictory in themselves' insofar as they are 'not shared by everyone', and proceed to develop Wigner's 1967 experiment that focuses on overcoming the isolation of the laboratory within which quantum systems with particles in particular spin states are entangled with particles observed outside the lab by another agent elsewhere. Thus, in challenging the famous superpositioning (within a box) of live/dead cat(s), the Schrödinger cat paradox,[41] Frauchiger and Renner underscore how the 'presumption that the validity of quantum theory extends to larger scales has remarkable consequences'.[42] By engaging multiple frame possibilities, Frauchiger and Renner perhaps enable a means of entertaining not one theory (whether rooted in Bohr's complementarity, Bohm's mechanics based on adapting Schrödinger's formula, or Bell's retreat to local beables, to name a few), but many. Their work clarifies how to frame the *choice of frames itself* as determining the outcome of the resolution of issues of how quantum theories and classical paradigms resolve problems of scale in broader attempts to apply quantum theory. This approach echoes Tanabe's own realization that modern physics is dialectical on multiple levels; however, Frauchiger and Renner situate not only binary dialectical opposition but entire competing frames that form their own quantum logic of complex overlapping possibilities that can and cannot all be right. As a whole, these competing frames point to our ignorance and to the pending or limited nature of our understanding of quantum phenomena.

If we understand a key problem to be not one of spatially defined subjects and objects as being still caught up in Heisenberg's uncertainty principle or Bohr's complementarity, but rather one of finding it impossible at the subatomic level to resolve waves and particles at once, then the constructability of 'space' *as* a contested realm of competing frames becomes not a production of space so much as a postulation of probable frames for such postulations, which now appears natural as such within quantum thought. Along these lines of thought, a quantum intuition has yet to emerge, but it could.

The problem here is that a rather literal becoming is asserted based on a faith in a macro-physicality hypothetically conforming to an unknowable becoming of infinitesimal parts. It is what Reza Maleeh and Parisa Amani emphasize to be a form of 'weak objectivity' as the result of quantum experiments, namely an objectivity that is 'characterized by intersubjectivity' promoted by Bohr in lieu of 'pre-existing properties'.[43] Rather than accept this shift, one may emphasize the growing research in the foundations of quantum physics that situates the Copenhagen Interpretation as but one contending frame. This accepts that these are frames moving in the direction of attempting to overcome this unknowability as a mere reification of poststructuralist preferences for infinite complexity (which quantum physics writ large, scaled up a working ontology, would certainly entail). Nevertheless, without a means of clarifying how to account for ethical questions in refashioning our intuitions as we engage the world at a social scale, the quantum revolution cannot begin to enter culture as a contending new ontology.

How to define what is real – the unvisualizable mathematical realism of formal matrix mechanics based on observable qualities attributed to particles in Heisenberg or the more visualizable realism tethered to the manifest world of classical Newtonian physics through Bohr's adaption of Schrödinger's wave formula as wave realism – is a thorny issue. Slobodan Perovic summarizes how although Heisenberg's approach renounced 'individuality that characterizes classical particles', it 'was based on the emphasis on *discrete properties* of the observed phenomena, such as the occurrence of spectral lines of different intensities',[44] which suit representation in matrices. These debates in the 1920s set the parameters for subsequent experiments and competing theories; however, they have yet to reflect upon the current epistemic horizon, which in the form of database matrices form an obvious example of a non-visualizable real, one not bound to spatial embodiment yet defining us. An alternate model of the self may emerge based on data gathered through credit card records, search engine or social media profiles that are not a product of sensory engagement but rather a compiled malleable matrix of datasets that tip the argument away from a spatio-temporal definition of manifest relativistic or particle-wave superpositioning of possible states. Will such alternatives to classical manifest world ontologies emerge as a new quantum intuition, as an archetypal model for all being, and a matrixial intuition of one's self as traces of interaction not predicated primarily on one's own mental and sensory experience, but rather matrices predicated on shifting frames of engagement and connectivity to the world?

Alexander Wendt explores the use of '"shared" language [as] refer[ring] to a *superposition* state with which participants are entangled and thus [are] available non-locally to all'[45] and 'shared language entangles people through semantic non-locality'.[46] Complex ranges of possible life choices are objectively situated in matrices that can be used to suggest a new form of intuition and potential ranges of *a priori* frames[47] inspired by quantum mechanics. To focus, for instance, on 'touch' with respect to how the body organizes information is fairly compatible with classical physics; moreover, it remains bound to 'inclusion and intimacy (or exclusion and disdain)' in situating touch.[48] I read this approach as reflecting an intuitive sense that touch invites a reciprocal engagement (consciously analogous to Karen Barad's agential realism)[49] and forming 1) a spatio-temporal locus; and 2) a minimal spatio-temporal representable point of interaction that seems intuitively close to a quantum particle/wave system (usually consisting of one particle/wave measurement in a given model set-up to account for a conjugate pair of variables). However, the technology does not exist for experimentally realizing quantum touch or its interactive agencies at macrological scales; such arguments must proceed entirely by analogy, in speculation concerning the larger, scaled-up applicability of quantum mechanics to macrological and even social situations. If there is argument by analogy, it seems it should explore how this new intuition will replace classical and Kantian intuition. Could a solution be based not on an elegant reduction to the paradoxes of complementarity but rather on dialectics out of nothingness in Kyoto School philosophy to amend the space-time coordination of the world in classical Newtonian physics?

Questions for further discussion

1. As Kant situated *a priori* transcendental categories and human intuition in light of Newtonian physics, Nishida turns to Bridgman and others for support, while Tanabe sees a new critical mode deriving especially from Minkowski and Dirac; however, Tosaka appears to defer (less critically) to the officially sanctioned Communist Party line or Soviet positions. How can one justify each position in terms of such arguments?
2. In light of the discovery of the Higgs boson and Higgs fields – known through what emerges when elementary particles are sped up to near the speed of light and strike together in the Large Hadron Collider (LHC) – we now can speculate that the void of space is even less empty than we thought: known matter is just the tip of icebergs in a cosmic sea awash with free elementary particles just waiting to interact. Do the approaches of any of these three philosophers offer any particularly helpful insight in light of the role of nothingness as a starting point for Kyoto School thought? How might Nishida's discussion of *basho*, Tanabe's focus on Dirac, or Tosaka's discussion of Kantian space help situate Higgs fields?
3. What sort of update of the concept of *basho* would you propose in our age of general scientific challenges? Can there be a *basho* in an age of the internet, when we attend so much to virtual co-presences?
4. What insights from the emergence of the Standard Model of elementary particles and the universality of symmetries of charges (and other qualities) may be applied to extend Kyoto School approaches to situating the observer or various approaches to dialectics?
5. What implications for how we situate free will may be made from the discussions of Nishida, Tanabe and Tosaka in light of developments in physics today?
6. Looking back on the work of these three philosophers, could Nishida be said to use the word 'world' more than 'nature', which appears central to Tanabe and Tosaka? If so, why or why not?

References

Allori, Valia, 'Primitive ontology and the structure of fundamental physical theories', In: Alyssa Ney and David Z. Albert (eds), *The Wave Function: Essays on the Metaphysics of Quantum Mechanics*, 58–75, Oxford: Oxford University Press, 2013.

Allori, Valia, 'Quantum mechanics and paradigm shifts', *Topoi* 34, no. 2 (2015): 313–23.

Ananthaswamy, Anil, 'New Quantum Paradox Clarifies Where Our Views of Reality Go Wrong', *Quanta Magazine*, 3 December 2018.

Barad, Karen, *Meeting the Universe Halfway: Quantum Physics and the Entanglement of Matter and Meaning*, Durham, NC: Duke University Press, 2007.
Bell, J. S., *Speakable and Unspeakable in Quantum Mechanics*, Cambridge: Cambridge University Press, 1987.
Bohr, Niels, 'On the Notions of Causality and Complementarity.' *Science*, New Series, 111, no. 2873 (20 January 1950): 51–4.
Camilleri, Kristian, *Heisenberg and the Interpretation of Quantum Mechanics: The Physicist as Philosopher*, Cambridge: Cambridge University Press, 2009.
Coole, Diana H. and Samantha Frost, *New Materialisms: Ontology, Agency, and Politics*, Durham, NC, and London: Duke University Press, 2010.
Frauchiger, Daniela and Renato Renner, 'Quantum theory cannot consistently describe the use of itself', *Nature Communications* 9, no. 3711 (2018): 1–10.
Hegel, Georg W. and George di Giovanni, *The Science of Logic*, Cambridge: Cambridge University Press, 2010.
Hegel, Georg W. and M. J. Petry, *Philosophy of Nature*, vol. 2, London: George Allen and Unwin, 1970.
Heisig, James W., *Philosophers of Nothingness: An Essay on the Kyoto School*, Honolulu: University of Hawai'i Press, 2001.
Kant, Immanuel, Paul Guyer and Allen W. Wood, *Critique of Pure Reason*, Cambridge: Cambridge University Press, 1998.
Kopf, Gereon, 'Between Identity and Difference: Three Ways of Reading Nishida's Non-Dualism', *Japanese Journal of Religious Studies* 31, no. 1 (2004): 73–103.
Kozyra, Agnieszka, 'Nishida Kitarō's Philosophy of Absolute Nothingness (*Zettaimu no Tetsugaku*) and Modern Theoretical Physics', *Philosophy East and West* 68, no. 2 (2018): 423–46.
Krummel, John W., *Nishida Kitarō's Chiasmatic Chorology: Place of Dialectic, Dialectic of Place*, Bloomington: Indiana University Press, 2015.
Maleeh, Reza and Parisa Amani. 'Pragmatism, Bohr, and the Copenhagen Interpretation of Quantum Mechanics', *International Studies in the Philosophy of Science* 27, no. 4 (2013): 353–67, DOI: 10.1080/02698595.2013.868182.
Maraldo, John C., 'Nishida Kitarō', in Edward N. Zalta (ed.), *The Stanford Encyclopedia of Philosophy* (Winter 2019 Edition), https://plato.stanford.edu/archives/win2019/entries/nishida-kitaro/.
Maudlin, Tim, *Philosophy of Physics: Quantum Theory*, Princeton, NJ: Princeton University Press, 2019.
Mellström, Ulf, 'From a Hegemonic Politics of Masculinity to an Ontological Politics of Intimacy and Vulnerability? Ways of Imagining through Karen Barad's Work', *Rhizomes: Cultural Studies in Emerging Knowledge* 30 (2016), https://doi.org/10.20415/rhiz/030.e07.
Murdoch, Dugald, *Niels Bohr's Philosophy of Physics*, Cambridge: Cambridge University Press, 1987.
Nishida, Kitarō, *Last Writings: Nothingness and the Religious Worldview*, trans. David A. Dilworth, Honolulu: University of Hawai'i Press, 1987.
Nishida, Kitarō, *Nishida Kitarō zenshū*, 19 volumes, Tokyo: Iwanami Shoten, 1989. Abbr. NKz.

Perovic, Slobodan, 'Why were matrix mechanics and wave mechanics considered equivalent?', *Studies in History and Philosophy of Modern Physics* 39 (2008): 444–61.

Schultz, Lucy, 'Nishida Kitarō, G. W. F. Hegel, and the Pursuit of the Concrete: A Dialectic of Dialectics', *Philosophy East and West* 62, no. 3 (July 2012): 319–38.

Tanabe, Hajime, *Tanabe Hajime zenshū*, 15 volumes, Tokyo: Chikuma Shobō, 1964. Abbr. THz.

Wendt, Alexander, *Quantum Mind and Social Science: Unifying Physical and Social Ontology*, Cambridge: Cambridge University Press, 2015.

Notes

1 Introduction: Relativity and Quantum Physics in the Kyoto School

1. Emmanuel Levinas, *Totality and Infinity: An Essay on Exteriority* (Pittsburgh: Duquesne University Press, 1969), 191.
2. Levinas, *Totality and Infinity*, 189.
3. See Yoko Arisaka, 'Beyond "East and West": Nishida's Universalism and Postcolonial Critique', *The Review of Politics* 59, no. 3 (1997): 541–60.
4. Michael Friedman, 'Newton and Kant on Absolute Space: From Theology to Transcendental Philosophy', in Michel Bitbol, Pierre Kerszberg and Jean Petitot (eds), *Constituting Objectivity: Transcendental Perspectives on Modern Physics*, 35–50 (Dordrecht: Springer, 2009).
5. Such philosophers include Fukuzawa Yukichi 福澤 諭吉 (1835–1901), Nishi Amane 西周 (1829–97) and Inoue Tetsujirō 井上 哲次郎 (1855–1944).
6. Niels Bohr scholar, Katsumori Makoto, emphasized this point to me in conversation.
7. This is partly because many of the first Western scholars interested in the Kyoto School happened to be religious people engaged in interreligious dialogue or philosophers interested in religion.
8. On Tanabe's relationship with Heidegger, see especially James W. Heisig, *Much Ado about Nothingness: Essays on Nishida and Tanabe* (Nagoya, Japan: Nanzan Institute for Religion and Culture, 2015); and John W. Krummel, 'Chōra in Heidegger and Nishida', *Studia Phænomenologica* 16 (2016): 489–518. Nishida was influenced by Heidegger through his students who had studied in Germany. See also Curtis A. Rigsby, 'Nishida on Heidegger', *Continental Philosophy Review* 42 (2010): 511–53; and Toru Tani, 'Inquiry into the I, Disclosedness, and Self-consciousness: Husserl, Heidegger, Nishida', *Continental Philosophy Review* 31 (1998): 239–53.
9. See THz, 6–7.
10. The most extensive exploration of Tanabe in this regard in English is Naoki Sakai, 'Subject and Substratum: On Japanese Imperial Nationalism', *Cultural Studies* 14, no. 3–4 (2000): 462–530.
11. For an introduction to wartime politics and the Kyoto School, covering postwar debates and competing positions and interpretations, see John C. Maraldo, 'The War over the Kyoto School', *Monumenta Nipponica* 61, no. 3 (2006): 375–406.
12. See Jacynthe Tremblay, 'L'influence du concept de complémentarité dans la philosophie du dernier Nishida', *European Journal of Japanese Philosophy* 3 (2018): 57–77.
13. See Eikoh Simao, 'Some Aspects of Japanese Science, 1868–1945', *Annals of Science* 46 (1989): 69–91.

2 Nishida Philosophy, Place, Field and Quantum Phenomena

1. See Michiko Yusa, 'Mathematics or Philosophy? (1886–1891)', in *Zen and Philosophy: An Intellectual Biography of Nishida Kitaro* (Honolulu: University of Hawai'i Press, 2002), Ch. 2, esp. 29.
2. Minkowski developed some of the most important thought in physics before and after Einstein's theories of general and special relativity, including four-dimensional space-time, in which time is given its own axis coordinated with three-dimensional space, and the idea that the trajectory of objects can be traced on the resulting 'world-line'.
3. See John C. Maraldo, *Japanese Philosophy in the Making 2: Borderline Interrogations* (Nagoya: Chisokudō Publications, 2019), 403–62.
4. NKz1, 6–7, translation from Fujita, 2018, 14.
5. Matteo Cestari, 'The Knowing Body: Nishida's Philosophy of Active Intuition [*Kōiteki Chokkan*]', *Eastern Buddhist*, New Series, 31, no. 2 (1998): 202–3.
6. NKz4, 49.
7. Although in usage today (within and beyond philosophy), *hyōgen* 表現 means 'expression', but in philosophy during Nishida's writing of this piece, it could also serve as another word along the lines of 'representation' or 'presentation', so that one may also consider translating this phrase as '(re)presentation of will'. Moreover, according to the *Iwanami Dictionary of Philosophy* (1922), the verb *hyōgen suru* is equivalent to 'to represent' in English and '*Abbilden*' (to depict, to show) in German. See Wakichi Miyamoto, *Iwanami tetsugaku jiten* (Iwanami's Dictionary of Philosophy) (Tokyo: Iwanami Shoten, 1922), 778. However, in the end, 'expression' is almost always found to be the best choice for conveying the distinct Kyoto School focus on dialectics that in both Nishida and Tanabe situate the subject and world in terms of various modes of expression with various forms of agency involved. Many thanks to John Maraldo for offering ideas regarding this translation problem in consultation via personal correspondence. Also compare his citation of a similar passage in a later essay by Nishida in Maraldo, *Japanese Philosophy in the Making 2*, 460, n. 61.
8. NKz4, 59.
9. 'Butsuri-genshō no haigo ni aru mono' (first published in 1924) was included in the collection *From the Acting to the Seeing* (*Hataraku mono kara miru mono he*, 1927). Translation based on NKz4, 48–75 (old edition).
10. Nishida seems to meld Kantian *a priori* transcendental concepts with a Hegelian sense of disunity and innate struggle or dialectics inherent to each situation. *Idea* seems to store or sublate (transcend, move beyond while building upon) sensory contents in a more dynamic and immanent Neo-Kantianism that he develops independently.
11. Nishida links the intuition of colour to a universality that contains the particular colours. This may be understood as reflecting his inheritance of Aristotelian aspects of form (*hylomorphism*,). As John W. M. Krummel writes, 'Nishida collapses this hylomorphism. He finds the root of this form-matter unity in the will's living experience understood in terms of *basho*.' See Kitarō Nishida, *Place and Dialectic: Two Essays by Nishida Kitarō*,

trans. John W. M. Krummel and Shigenori Nagatomo (Oxford: Oxford University Press, 2012), 20.
12 Here Nishida uses 'idea' in the Platonic sense of the objective form of something. Concept, by contrast, then is not bound up with a particular but rather something more incidental or general.
13 Cf. Leibniz's monadology.
14 A good introduction to the importance of William James for Nishida, who had practised Zen Buddhist meditation for years and found James's concept of pure experience a common nexus with Western philosophy by which his own philosophy could be constructed, is Joel W. Krueger, 'The Varieties of Pure Experience: William James and Kitaro Nishida on Consciousness and Embodiment', *William James Studies* 1, https://williamjamesstudies.org/the-varieties-of-pure-experience-william-james-and-kitaro-nishida-on-consciousness-and-embodiment/.
15 Here Nishida is certainly adapting Aristotle's contrasting of universal ('of a whole') as distinct from the particular in a logical proposition such as 'Cleopatra is a woman', whereby a particular is part of a general universal category, though the inverse, 'a woman is Cleopatra', makes no sense. According to Robin Smith, 'This distinction is not simply a matter of grammatical function. We can readily enough construct a sentence with "Socrates" as its grammatical predicate: "The person sitting down is Socrates". Aristotle, however, does not consider this a genuine predication. He calls it instead a merely **accidental** or **incidental** (*kata sumbebêkos*) predication. Such sentences are, for him, dependent for their truth values on other genuine predications (in this case, "Socrates is sitting down"). Consequently, predication for Aristotle is as much a matter of metaphysics as a matter of grammar. The reason that the term *Socrates* is an individual term and not a universal is that the entity which it designates is an individual, not a universal.' See Robin Smith, 'Aristotle's Logic', in Edward N. Zalta (ed.), *The Stanford Encyclopedia of Philosophy* (Summer 2019 Edition), https://plato.stanford.edu/archives/sum2019/entries/aristotle-logic/. Much Nishida scholarship has treated *shugo* as 'grammatical subject'. It may also be rendered 'logical subject'.
16 Note the implication of a Hegelian telos here, a convergence of sentiment and quasi-spontaneous coordination of public spirit.
17 Here he reflects dialectical relations found throughout Hegel, including his *Phenomenology of Spirit* and *Science of Logic*.
18 Note how Nishida shifts from the socially situated human *subject* (*shutai*) to the propositional subject (*shugo*).
19 This sentence shows how Nishida Philosophy, like other philosophers of his era (notably Whitehead and William James), speaks to post-human issues of human–non-human relationality. Nishida's analysis of how qualities are distinguished preserves a role for will that provides an alternative to the more concept-oriented systems derived from Euro-American philosophy, as he presents a method of inquiry that constantly deconstructs assumptions while attending to a modelling that demands a dynamic pluralism that accounts for the human subject, accounts for speculative methods of presentation in philosophical form through various adaptations of propositional logic to physical situations and states, and that accounts for non-human actors and the production of qualities as non-essential

attributes. In other words, these texts long ignored in English scholarship speak to a sensibility emerging both in Nishida's day and our own.
20 Note how this dual attention to Kantian questions of how objects are mediated speaks to issues in quantum mechanics only introduced later in the philosophy of science in general and in Nishida's writings.
21 Here the earlier mention of Leibniz and god reappears in this allusion to the world as monad. Nishida's own thought increasingly complicates the relation of parts and wholes in terms of unity and discrete individuals.
22 Nishida appears to point to the spatiality and temporality of objects with respect not to placement within a larger configuration of space-time (or space and time when referencing pre-relativity models), but the spatio-temporality within them as apart from sensorial qualities.
23 Here the influence of Nietzsche and Schopenhauer appears, linking will to expressions of power and representation.
24 This phraseology may suggest to readers today much later writings by Henri Lefebvre, especially his tome translated into English as *The Production of Space* (Oxford: Blackwell Publishing, 1991).
25 Though some readers may object to '*ketsugō suru*' (literally 'combine') being translated as 'synthesize', the English rendering would be awkward and forced, simply not as clear.
26 The phrase translated as 'birth and death' is a Buddhist term for the cycles of life. In the subsequent passages one sees the originality of Nishida's thought in its assertion of a Buddhist frame based on a finite present where consciousness appears with various configurations of immanent force in lieu of Hegelian time, with its signature sublation of dialectical conflicts as space and time form one type of dialectical relationship diverging from the sense of space and time as products of Kantian intuition through the instantiation of *a priori* transcendental categories of space and time.
27 Later, as quantum physics develops new approaches to space and time, the role of abstract space will be elevated in the context of atomic experiments and theories in the form of the Hilbert configuration or vector space not coincident with worldly coordinate space, creating extremely complex problems in the study of the foundations of physics.
28 Nishida often situates his subjects in such forms of interrelationship, but in his writings on physics he turns his attention to the formation of objectivity and the underlying principles posited by Kant and Hegel regarding space and time, and all the various thoughts emerging in his day and over the last century.
29 This dynamic relation of will and force to the production of space reflects a pre-quantum physics worldview, a Newtonian and more-or-less Euclidean (technically Minkowski's proto-relativity theory approach to space-time) view that maintains a conservation of mass and energy, which is one of the most fundamental principles of classical physics. It also implies a standard basis for causal relations within a whole universe only altered by relativity theory, but not fundamentally shaken as it will be by quantum mechanics, as discussed in subsequent chapters.
30 This assertion certainly reflects Nishida's reading of William James.
31 The importance of Fichte cannot be overstated here, since the collection of essays within which this chapter is included focuses on this theme of the act as the foundation of the knowable.

32 Again one sees a defined turn away from a Kantian mentally based *a priori* apparatus based in part on transcendental categories and a turn toward German idealism, both Fichte's fact-act and Hegelian dialectics.
33 Here we see Nishida build towards the construction of the best-known concept in his philosophy, the *basho*. The *basho* serves as a site grounding consciousness in a specific location, shifting questions of self and other as well as subject and object to a field of relations. In terms of physics, the question remains in Nishida studies, to what extent is this concept derivative of a Newtonian (and Kantian) perspective or reflect relativity theory and quantum theory? These questions are only now being broached in Nishida studies. See Tremblay, 'L'influence du concept de complémentarité dans la philosophie du dernier Nishida'.
34 One might consider how this approach differs from Hegelian sublation (*Aufhebung*), which sees a dialectical continuum as cancelling through overcoming and yet containing the past forgotten elements in the form of an unwieldy collection of logical steps. Here, Brentano and Nishida Philosophy in general focus on a moment of cognition within which a *field* of knowables is entertained *in relation to various parameters of recognition* that include relations to others, temporal and spatial proximity, remembered experiences (as 'oblique' presentations) and so forth.
35 Nishida refers to Oskar Kraus (1872–1942), author of *Franz Brentano: Zur Kenntnis seines Lebens und seiner Lehre, mit Beiträgen von Carl Stumpf und Edmund Husserl* (Munich: Beck, 1919), probably 40–1.
36 Note that this formulation differs from Hegelian or other dialectics in that it is a structural, permanent interplay of elements – time and space as mediated by a necessary observer – rather than the building of an artifice of spirit out of human struggle and overcoming.
37 See Immanuel Kant, *Critique of Pure Reason* (New York: Cambridge University Press, 1998), esp. 290–5.
38 From principles of mathematics to principles of dynamics.
39 Nishida is alluding to the Kantian *a priori* transcendental categories that depend on the foundational intuitions of time and space (as antinomies) as well as the intuition of qualities (as categories). He follows Heinrich Hertz, whose approach to dynamics dispensed with force in ways that anticipate modern physics.
40 Vectors describing the change of the path of a particle in four-dimensional space-time in Minkowski and special relativity.
41 Theodore Lipps cited in Robin Curtis, 'An Introduction to Einfühlung (*Empathy*)', trans. Richard George Elliott, *Art in Translation* 6, no. 4 (2014): 359.
42 If one wishes to find traces of Nishida's debt to D. T. Suzuki and Zen meditation practice (sitting Zen) in general, here one may speculate that he has situated similar self-reflection as a practice, a convention.
43 Compare Kant on space as an intuition based on *a priori* categories instantiated in perception.
44 This key point directly links suggestions of Hilbert coordinate space that stands apart from the mundane world we live in, and may form a foundation for consideration of later issues concerning quantum mechanics and questions of the relevancy of its particle-system scaled observations to the macrological world.

45 This discussion of Fichte I believe clarifies the role of the absolute in Nishida as a means of conceptualizing both opposition of subject and object and grounding of the self in a frame of self/non-self that provides a blank slate for the emergence of the world through the active self, rather than a self that reflects the world passively. As Scribner argues, Nishida diverges from Fichte's *Tathandlung* (fact-act) by situating the grounding of first (framing) principles not as determined beyond the subject by a process of codetermination of frame and subject. See F. Scott Scribner, 'Nishida's Fichte and the Resistance of Idealism', *International Journal for Field-Being* 1, no. 1 (Part 2, Article No. 13, 2001), www.iifb.org/ijfb/FSSScribner-2-13.
46 Literally, the awkward phrase 'in the object world [of] cognition' (*ninshiki taishōkai*).
47 For a full translation, see Kitarō Nishida, 'Basho', in *Place and Dialectic: Two Essays by Nishida Kitarō*, trans. John W. M. Krummel and Shigenori Nagatomo (Oxford: Oxford University Press, 2012), 49–102.
48 I try to distinguish 'world of objects' or 'object world' (*taishō-kai*) from the 'objective world' (*kyakkanteki(na)sekai*)) and 'realm of objectivity' (*kyakkan-kai*), since *taishō-kai* foregrounds the material world, the latter two a way of seeing that somehow privileges it.
49 This turn towards an abstraction *seeing* the self marks an important development in Nishida Philosophy, so important that it appears in the title of the collection of essays that also includes one on the far more influential concept of the *basho* (place, matrix, field, etc.): *From the Acting to the Seeing* (*Hataraku mono kara miru mono he*) (1927).
50 Plotinus (*c*. 204/5–270) was a Neoplatonic philosopher who influenced Nishida, contributing a sense of greater attention to questions concerning the hierarchies of matter and its organization in relation to the self, and the role of will. According to Longo and Taormina, 'Plotinus conceives the atomistic or generally corporealistic theories as positing a plurality of causes, as opposed to the Stoics, who posited one principle, the world-soul only, etc. He suggests that a plurality of kinds of principle is needed to explain higher-level order, and to avoid unacceptable determinism.' Angela Longo and Daniela Patrizia Taormina, *Plotinus and Epicurus: Matter, Perception, Pleasure* (Cambridge: Cambridge University Press, 2016), 174, n. 38.
51 Final causality is a translation from the German *Zweckmässigkeit*, which carries a sense of things maintaining definite purposes.
52 See introduction, on Plotinus.
53 Here one cannot help but notice similarities with Alain Badiou's emphasis on what is included or not, and how the engagement with the infinite is necessarily limited to sites of engagement (rather that absolute categories). In fact, Badiou's work is built explicitly on Cantorian set theory, also an influence on Nishida as he developed his Buddhist orientation toward infinity within continental philosophical discourse. See Matao Noda, 'East-West Synthesis in Kitarō Nishida', *Philosophy East and West* 4, no. 4 (1955): 349. See Alain Badiou, *Second Manifesto for Philosophy* (Cambridge and Malden, MA: Polity, 2011), esp. 111–15.
54 Here the translation is necessarily awkward, since the ideas conveyed required some relinquishing of commonplaces of thought in English usage, usage that is difficult to capture. The main point here seems to be that since the will is bound up in all discernment of motion, including stilled motion,

the action of the will too represents an infinite engagement in acts of unifying the infinite play of elements. Next, he will outline more specifically Neo-Kantian issues of how the structuration of the world remains invested in the physical world. This essay thus presents a Kantian groundwork for situating modern physics more directly in later essays.

55 The Buddhist phrase used here, *dōsei ichi'nyo*, is literally the unity, sameness or single suchness (*ichi'nyo*) of movement and stillness (*dōsei*). In the context of Nishida, it implies a convergence of a Buddhist-like deconstruction of assumptions about the misleading appearances and distinctions present to us in the world and modern physics's deconstruction of classical physics.

56 This direction in Nishida's thought is precisely what leads him to find affinity later with certain approaches to quantum physics as well as Bridgman's Operationalism, the focus of section 2.2 that follows. These approaches variously engage problems of locating subjects in light of modern physics, and for Nishida, using modern physics to justify his own arguments such as the above in building his philosophy.

57 Though Nishida may not have consciously alluded to Hegel's discussions of time, similarities in the approach to time exist. See the discussion in Chapter 5.

58 *Keiken kagaku*; NKz9, 223–304.

59 For instance, Bridgman writes that 'the measurable properties of electrons embrace some phenomena which we find convenient to describe in terms of the wave phenomena of ordinary experience, *in addition to the older and more familiar phenomena which we have satisfactorily dealt with in terms of a particle picture*'. P. W. Bridgman, 'Permanent Elements in the Flux of Present-Day Physics', *Science* 71 (1828/1930): 22 (emphasis added).

60 For a narrative overview of the discovery of the Higgs boson, see Sean Carroll, *The Particle at the End of the Universe: How the Hunt for the Higgs Boson Leads Us to the Edge of a New World* (New York: Dutton, 2012).

61 The only way Bridgman's approach to recreating a classical model for explaining experimental data from quantum mechanical operations applies would be to take the more helpful formula for estimating probable states for wave-particles such as Schrödinger's equation, which takes Planck's constant h as a means of recreating a still world (containing a space of an object). It is vaguely parallel to how Einstein invoked c, the (still controversial, in light of multiverse theory) speed of light to recover a semblance of Euclidean space in a Relativistic world.

62 See Agnieszka Kozyra, 'Nishida Kitarō's Philosophy of Absolute Nothingness (*Zettaimu no Tetsugaku*) and Modern Theoretical Physics', *Philosophy East and West* 68, no. 2 (2018): 425–6.

63 *Keiken kagaku* (originally appeared in the journal *Shisō* 207, August 1939). Translation based on NKz9, 223–304 (abridged), presented as Chapter 4 of *Collection of Philosophical Writings: Volume 3*. Since from Nishida's first book, *Zen no kenkyū* (*An Inquiry into the Good*), he has pursued a greater understanding of the experience of a person in situating ontological being and epistemological meaning, so that the 'empirical' in the title being the same two characters for 'experience' should be noted, as it connotes a 'science of experience'.

64 The two works by P. W. Bridgman that I refer to are *The Logic of Modern Physics* and *The Nature of Physical Theory* (Nishida's note).

65 Of course, here Nishida presents his exploration of how we frame phenomena in light of how frames themselves provide *a priori* parameters for situating objects in relation to processes, that is, neo-Kantian questions of the unknowable thing-in-itself and how we still attempt to approach things.
66 It should be noted that the word for experience (*keiken* 経験) has as an adjective of the same root found in empiricism. Also, that Nishida did not use a related word that might be expected here – *jikken* (実験) or 'experiment' – is significant, for it is essential to the progressive shift of his engagement with modern physics discussion to discussion of his own philosophical system, which is rendered in schematic form later in this chapter.
67 By 'our' approach to length, of course, Nishida is referring to classical Newtonian physics, which still largely forms our common sense concerning measurement of objects in space.
68 Note that in Japanese the word translated here as 'fields' (*ba*) is a key term in Nishida's philosophy, the philosophy of place or field. Here we can see the importance of Bridgman and modern physics in general to Nishida's defence and development of his own philosophy.
69 The omitted paragraph argues for the importance of measurement based on first-hand experience. This point reinforces Nishida's commitment to situating the self in the investigation of empirical knowledge.
70 Nishida's insertion in English, without parentheses.
71 This introduction of previous experiences as elements of knowledge commensurate with the operationalism of physics experiments again ties in Nishida's own philosophy to Bridgman's work on physics.
72 On the one hand, here Nishida reflects not only Bridgman but a far more influential physicist within quantum physics, one that may also have been influenced by Bridgman: Niels Bohr, who was famous for his adherence to the maintenance of a place for natural language description as well as classical physical presence of person, lab and so on (as part of his response to Einstein, Podolsky and Rosen, the EPR paper). On the other hand, Nishida is adhering to his lifelong philosophical investigation of the nature of experience and sites of knowing. See Dugald Murdoch, *Niels Bohr's Philosophy of Physics* (Cambridge: Cambridge University Press, 1987), esp. 148.
73 The omitted paragraph merely describes the limits of Euclidean mathematics and the necessity of conducting actual measurements based on empirical knowledge. It argues that 'no matter how refined one's process of measurement becomes, it will never approach [the fullness of] Euclidean geometry'.
74 While Nishida is repeating Bridgman's argument in his own words, his wording and style reflects his own philosophical interests and agenda. Here too the focus on 'site' (*basho*) appears as a central problem applicable to the development of his own thought.
75 This sentence really brings into focus both the problem of distinguishing parts and wholes within his philosophical approach that depends on a subject situated as a human, and the reason Nishida turned to modern physics for support of his approach, assuming that the natural sciences would provide a firmer ground than speculative thought, no matter how

carefully articulated and developed in recursive steps, as we see Nishida tending to do.
76 Although a minor point, note how the word 'experience' substitutes for 'experiments' or 'empirical proof'. Nishida is blurring the laboratory-controlled conditions attendant to most physics experiments in order to situate his own philosophical system as universal and supported by modern physics.
77 The omitted pages (231–7) are fascinating speculation on the relation of mathematics to language in a world within which spatialization itself becomes problematic, upon the advent of atomic physics, which disrupts classical concepts such as unified space as well as causality predicated on forces quantified in spatio-temporal frames of relation.
78 Bridgman, 'Permanent Elements', 20–1.
79 Bridgman, 'Permanent Elements', 21.
80 Bridgman, 'Permanent Elements', 22.
81 Here one clearly sees Nishida linking Operationalism to the human subject so as to historicize what has been called an idealist construct: the logic of the topos (*basho no ronri*), the centre of Nishida Philosophy.
82 *Gijutsu* may also be translated as 'technology' or 'art', as in skill. The English word 'technology' has a more objective connotation, and Nishida associates *gijutsu* with subjective skills (rather than an image of robots on an assembly line that one may imagine today).
83 The omitted paragraph (238–9) is somewhat repetitive.
84 The omitted paragraph (240–1) is somewhat repetitive.
85 It should be clarified that the word for 'pictured' in this passage is *utsusu* (映す), also translatable as 'reflected', 'projected' or even 'imagined'. English apparently lacks a word with the sense of both projecting what is of the mind and of the world in this sense, since 'picture' and 'imagine' are biased towards an inner action; however, Nishida is seeking to emphasize the act of making (*poiesis*) or rendering of the world, so this bias works well in this regard. 'Depict' or 'render' would introduce a sense of drawing (*egaku*) or presenting (*utsushi-dasu*) that is perhaps too biased to the objective.
86 In this sentence, the first 'individual' is human (*kojin*) while the second is a more general and unspecified philosophical distinction, also rendered 'individual thing' (*kobutsu*).
87 This discussion of instinct certainly derives from reading not only Bridgman here but Bergson, whose vitalist account of the origins of life as self-organizing in relation to other beings invokes a spectrum of instinct and intelligence. See Leonard Lawlor and Valentine Moulard Leonard, 'Henri Bergson', in Edward N. Zalta (ed.), *The Stanford Encyclopedia of Philosophy* (Summer 2016 Edition), https://plato.stanford.edu/archives/sum2016/entries/bergson/.
88 The omitted paragraph (244–5) draws Bridgman into the Bergsonian discussion of instinct and reason.
89 In the sense of the German word from which this word derives, *Zweckmässigkeit*, which has a sense of appropriateness, expediency or efficacy.
90 The omitted paragraph (247–8) discusses Leibniz's monad, not necessary for understanding the argument.

91 The possibilities for translating this most important of Nishida's terms, *basho*, include both Greek *topos* (with its sense of physical place) and *chōra* (matrix, with its sense of necessary frame), as well as field (in the sense of post-classical modern physics). See John W. Krummel, *Nishida Kitarō's Chiasmatic Chorology: Place of Dialectic, Dialectic of Place* (Bloomington: Indiana University Press, 2015).

92 I seem to differ from most existing renderings in exploring translating *hyōgen* as 'representation' rather than 'expression', following not only the descriptions in philosophy dictionaries but also how it better expresses Nishida's sense of the interaction of world and self here and elsewhere with more objective remove. Ultimately, I agree with 'expression', while we can learn from struggling with this issue.

93 In the omitted section, Nishida argues that our techniques and technologies becomes analogous to instinct with respect to how our bodies function operationally over time and in relative space.

94 Nishida read and engaged the work of American objective idealist philosopher Josiah Royce (1855–1916), whose two-volume *The World and the Individual* (1899, 1901) maintained a central place for the object world and the mediating self, exploring the relationship of the one and the many at length.

95 Note here how translating *jiko-hyōgen* as self-expressive would suggest a personification of 'historical space' (seen here more or less as a point in moving time in a Minkowski world-line) and entirely miss the balanced, mutually-negating aspect of Nishida Philosophy and reduce it to a form of inflected subject-bound expression rather than co-produced (subjective and objective) representation. However, part of appreciating Nishida's innovative thought depends on some attempt to approximate this direction as 'self-expressive' (or perhaps, more awkwardly, 'self-presenting').

96 From this point Nishida returns to the work of Kurt Lewin, first introduced early in an untranslated passage. He refers to two works: Kurt Lewin, *Principles of Topological Psychology*, trans. Fritz Heider and Grace M. Heider (New York and London: McGraw-Hill Book Company, 1936); and Kurt Lewin, *A Dynamic Theory of Personality*, trans. Donald K. Adams and Karl E. Zener (New York and London: McGraw-Hill Book Company, 1935).

97 This statement of situated causality (again) reveals Nishida's adherence to laws of classical physics despite the rise of relativity and quantum mechanics. Bridgman's limited interpretations of the new physics in this regard formed an imperfect source for engaging key issues in such new terms.

98 Here is one of Nishida's rather rare (until his very late work) references to a specifically Buddhist term, 'ultimate truth', with Buddhist connotations of a world of cause and effect existing in a zero-sum model of actions that entail responsibility (karmic consequences).

99 See Box 2.3.

100 Of course this alludes to the thoroughly Hegelian sublation (*aufgeben*) of negated elements in a dialectical process.

101 Put forth by Descartes, Leibniz and others, it is literally a 'universal' 'science' based on mathematical modelling, with the premise that such an approach

would provide superior access to truth in contrast with natural language alone.
102 Again, one may note the precise use of 'representation' (*hyōgen*) here serving 'as' a negation in a Nishida Philosophy dialectic. Though contemporary usage of *hyōgen* 'expression', rendered 'representation', indeed clarifies Nishida's usage and maintains the sense contemporary to his day. See '*hyōgen suru*' in Wakichi Miyamoto, *Iwanami tetsugaku jiten* (Iwanami's Dictionary of Philosophy) (Tokyo: Iwanami Shoten, 1922), 778.
103 Nishida carefully chooses *hataraku* here not simply to build upon his earlier work published in a volume with this title, but because it reinforces Bridgman's approach to problems of space-time in physics in terms of Operationalism, which in turn supports Nishida Philosophy as a philosophy rooted in physical reality and by extension historical reality (an issue he engaged since related criticism arose much earlier).
104 Again, these are precisely what seem to have drawn Bohr to Bridgman's Operationalism. How does this shape Nishida's thought?
105 Copy theory is based on the problematic idea that knowledge may be construed simply by modelling the world and phenomena in reproductions, in images apart from the things in themselves. See Frederick D. Wilhelmsen, *Man's Knowledge of Reality: An Introduction to Thomistic Epistemology* (New York: Prentice-Hall, 1956), 77.
106 It should be emphasized that this modelling suggests a human form of modelling in conjunction with our bodies.
107 The three instalments of these schematics are found in NKz8, 219–66 (1935), 572–9 (1937) and NKz9, 305–35 (1939).
108 'Schematic explanations' (1935), NKz8: 219–66.
109 See NKz7, 213–15.
110 NKz8, 572–89.
111 Lewin, *Principles of Topological Psychology*, 73.
112 NKz9, 255.
113 See NKz8, 224.
114 See NKz7, 213.
115 NKz8, 219.
116 NKz8, 219.
117 See NKz7, 213.
118 See NKz8, 232.
119 See NKz7, 213.
120 See NKz8, esp. 219–66.
121 See NKz9, 305.
122 NKz9, 307.
123 NKz9, 307.
124 NKz8, 229.
125 NKz8, 229–30.
126 NKz8, 230–1.
127 See Murdoch, *Niels Bohr's Philosophy of Physics*, 148–9.
128 See Tremblay, 'L'influence du concept de complémentarité dans la philosophie du dernier Nishida', 65.

3 Mediation in Tanabe's Dialectical Vision of Competing Fields within Physics

1. See Chapter 2.
2. Hideki Mine, *Nishida tetsugaku to Tanabe tetsugaku no taiketsu: basho no ronri to benshōho* (Kyōto-shi: Mineruva Shobo, 2012).
3. THz5, 269.
4. Karen Barad, *Meeting the Universe Halfway: Quantum Physics and the Entanglement of Matter and Meaning* (Durham, NC: Duke University Press, 2007), 249–50.
5. THz5, 269
6. THz5, 270
7. THz5, 270
8. THz5, 283
9. THz5, 284.
10. THz14, 452–69. Note that this work does not appear in all editions, and is listed as Appendix 2 (補遺2).
11. Bohr, *Atomic Theory and the Description of Nature*, 2 (Tanabe's note, referring to a Japanese translation of this 1931 work.)
12. 'System', as used in modern physics, refers usually to contained coordinate systems in isolation, with no assumption of a real or situated space.
13. Here Tanabe is surely referring to the transcendentalism found in Kant and Husserl, not the American transcendentalists such as Emerson and Thoreau. In Kant and Neo-Kantianism, the transcendental is integral to perception and experience as it takes the form of *a priori* categories by which things are recognized. Thus ontological and epistemological forms are mutually implicated in the process of empirical and transcendental knowing of the world. Contemporary philosophy of physics has been moving in some works towards a more detailed realization of such Kantian approaches to how matter can be grasped in patterns. See Jan Faye and Henry J. Folse, *Niels Bohr and the Philosophy of Physics: Twenty-first-century Perspectives* (London: Bloomsbury Academic, 2017), esp. chapters 2, 3 and 7.
14. To be clear, what Tanabe is saying is that measurement itself cannot leapfrog the complications of mutually determining locations in time and space relatively by relying on the old common denominator of the ubiquitous aether; instead, classical physics must now account for such phenomena in terms of Einstein's equations placing the constant of the speed of light as the defining saving characteristic of systems of spatio-temporal reference.
15. Tanabe surely is using 'commutation point' in the sense used in physics and mathematics, where it indicates a point of exchange between two systems of spatial reference. In quantum physics, such relations are situated by the use of various operators (necessary in a world without aether, nor absolute space or time).
16. One may compare Tanabe's use of 'common sense' with that elaborated by Tosaka Jun in Chapter 4.
17. Though writing in plain and somewhat jargon-free Japanese here, Tanabe seems to be alluding to a structure of sublation (*Aufgehoben*), which in Hegelian dialectics indicates a negation-incorporation.

18 Minkowski, *Space and Time*, 1; Bohr, *Atomic Theory and the Description of Nature*, 2 (Tanabe's note, referencing a Japanese translation.)
19 Sugimura Yasuhiko, 'Genshōgaku kara "shakai sonzai" e – 1934-nen no Tanabe to Levinasu' (From phenomenology to social existence: Tanabe and Levinas in 1934), *Shūkyō-gaku kenkyūshitsu kiyō* 14 (2017): 3–41, explores how Tanabe situates 'world', time and negation in relation to his writings on Heidegger and Levinasian issues in the opening chapter of his *The Logic of Species*, 'From Schematic "Time" to a Schematic "World"' (first published in 1932). In light of this work, Sugimura shows, Tanabe can be understood as avoiding Heidegger's 'schematic "time"' and promoting an 'schematic "world"' (24), an approach that demonstrates the continued influence of Nishida (by way of his work on *basho* and his dialectical situating of the subject in general). What may be added to this account is the importance of the origins of this usage in Tanabe due to his understanding and grounding of ontology in modern physics, as Minkowski's world is explicitly discussed in this 1932 piece (THz6, 21). In fact, some of this essay seems to cover similar ground with the 1937 essay, suggesting a cumulative process in Tanabe's writings on time, space and worlds that should be recognized as including physics.
20 Here one of the more well-known concepts in Minkowski, the 'world-line', is introduced as a variant to the other terms already mentioned.
21 The 'law of contradiction' applies to logic as a demand that a proposition cannot be both true and false. It also relates to the presence or not of properties in an object, so that modern physics challenges the nature of nature by disproving this apparent truism.
22 Planck, *The Worldview of the New Physics*, 18–19 (Tanabe's reference to a Japanese edition).
23 At this point, what Tanabe refers to is of course Bohr's first major contribution to quantum theory in his analysis of the atom, an approach that his and others' later work would displace in light of the 'new quantum theory', which resulted in part in the apparent paradoxes of complementarity in quantum phenomena (also see Box 3.10).
24 Tanabe is referring to Lorentz's modelling of relations between fields of reference after Einstein's special relativity theory would require additional considerations regarding the spatio-temporal symmetry of separated fields of reference now bereft of absolute space (or time).
25 The immensely influential precursor to modern physics, Ernst Mach (1838–1916), who was much admired by Tosaka.
26 De Broglie, *Collected Theses on Waves and Particles*, 73 (Tanabe's reference to a Japanese edition).
27 This question is still debated today, suggesting that Tanabe is being overly optimistic. Surely given the fast pace of developments in modern physics, his expectations may reflect the pace of new developments in the field at the time (a pace which would soon die down for some decades).
28 Schrödinger, 'Indeterminacy in Physics', 7 (Tanabe's reference to a Japanese edition).
29 Schrödinger, 'Indeterminacy in Physics', 15 (Tanabe's reference to a Japanese edition).
30 Bohr, *Atomic Theory and the Description of Nature*, 62–6 (Tanabe's reference to a Japanese edition).

31 Nishida too emphasized precisely these words attributed to Planck in an essay first published in 1915. See Nishida, Kitaro, *Intuition and Reflection in Self-Consciousness*, trans. Valdo H. Viglielmo, Takeuchi Toshinori and Joseph S. O'Leary (Albany, NY: SUNY Press, 1986), 58.
32 Tanabe's point seems to be that Spinoza's philosophical system is not simple, being complicated by the interrelation of affects and striving (*conatus*) in the formation of things in the world, not unlike the complications found in his own understanding of a dialectics of nature.
33 Here Tanabe is more explicitly engaging in debates initiated under the influence of Soviet debates on the relation of history and science, and especially the complications introduced by the new physics that were often simplified if not totally ignored by doctrinaire Marxists trying to make their thoughts on science adhere to the models proposed by famous works by Engels, Lenin, Deborin and others, all issues taken up in the materialist journal *Under the Banner of the New Science*. These issues are discussed in more detail in the next chapter.
34 Here, as often elsewhere in Tanabe, 'physical' is used to render the adjective *butsurigakuteki*, though it is usually a translation of *butsuriteki*. Indeed, *butsurigakuteki* has the sense of 'physical', but with an added sense of the 'physical as defined within physics' (which would be awkward to spell out in each instance). Strictly speaking then, 'physical' here means 'pertaining to the object of the science of physics, the object of which is the physical world'.
35 Here the grammar has been altered for comprehensibility. The phrase 'limits due to' is actually an adverbial expression of mathematical limits (*kyokugenteki ni* 極限的に) and what follows 'due to' is the object of 'express'.
36 In Minkowski's sense. See Box 3.3.
37 Tomio Nishikawa, 'On Tosaka Jun's Scientic-Technological Spirit', in Masakatsu Fujita, Robert Chapeskie and John W. Krummel, *The Philosophy of the Kyoto School* (Gateway East, Singapore: Springer, 2018), 93.
38 See Mine, *Nishida*, 139–40.
39 Heidegger alludes to and dismisses Einstein in a footnote in *Being and Time*, seeing his own conception of *Dasein* as creating the conditions for measurement in relativity. The dismissal seems unreasonable and possibly based on anti-Semitic bias. Tanabe certainly explored this lacuna left in Heidegger's major treatise. See Martin Heidegger, *Being and Time* (New York: HarperPerennial/Modern Thought, 2008), 499, n. 5.
40 Mine, *Nishida*, 259.
41 Mine, *Nishida*, 261.
42 Mine, *Nishida*, 263.
43 THz5, 318–27. This selection is the closing section of the chapter and book.
44 Indeed, imaginary (and real) numbers figure prominently in formulations attempting to comprehend observed behaviour at subatomic scales that can only be inferred mathematically due in part to the intrusive nature of measuring procedures themselves.
45 The above exposition is based on Dirac, *Principles of Quantum Mechanics*[II], 270–2 (likely near p. 275 in the 1967 edition); Dirac, *Theorie der Elektronen und Positronen. Die moderne Atomtheorie*, von Heisenberg, Schrodinger, Dirac, S. 37–45 (Tanabe's note).

46 On Aristotle's conceptualization of the 'not present-in', see Paul Studtmann, 'Aristotle's Categories', in Edward N. Zalta (ed.), *The Stanford Encyclopedia of Philosophy* (Fall 2018 Edition), https://plato.stanford.edu/archives/fall2018/entries/aristotle-categories/.
47 *Quantum Mechanics*, 7–13 (Tanabe's note). Tanabe seems to be citing Dirac's *The Principles of Quantum Mechanics*, though an earlier edition. See Chapter 1, 'The Principle of Superposition', in any edition.
48 *Quantum Mechanics*, 4 (Tanabe's note).
49 *Quantum Mechanics*, 17 (Tanabe's note).
50 Tanabe is referring to Kant's critical philosophy that incorporates concepts such as the transcendental unity in the process of perception of nature, not American transcendentalist philosophy.
51 Tanabe is not primarily alluding to classical Marxist philosophy but rather to Greek philosophy, especially Aristotelian concepts of matter.

4 Modern Physics, Space and Ideology in Tosaka Jun

1 See Kevin M. Doak 'Under the Banner of the New Science: History, Science, and the Problem of Particularity in Early Twentieth-Century Japan', *Philosophy East and West* 48, no. 2 (April 1998): 232–56.
2 Nishikawa, 'On Tosaka Jun's Scientific-Technological Spirit', 94.
3 See Tosaka Jun, *Tosaka Jun: A Critical Reader* (Ithaca, NY: East Asia Program, Cornell University, 2013).
4 Nishikawa, 'On Tosaka Jun's Scientific-Technological Spirit', 98.
5 See Doak, 'Under the Banner'.
6 TJz3, 239–66. See also, for more translated material from this work, Tosaka Jun, 'On Space', in *Tosaka Jun: A Critical Reader*, 17–35.
7 TJz3, 239.
8 See Nishikawa, 'On Tosaka Jun's Scientific-Technological Spirit', 94.
9 TJz3, 245.
10 TJz3, 263.
11 TJz3, 245.
12 Henri Lefebvre, *The Production of Space* (Oxford: Blackwell Publishing, 1991).
13 Lefebvre, *The Production of Space*, 45.
14 Lefebvre, *The Production of Space*, 12.
15 Lefebvre, *The Production of Space*, 7. Lefebvre's work is fascinating and illuminating, but in issues concerning the relationship of space and representation, representation and space. Lefebvre capitulates to the uncompelling notion of the 'in view' imposing 'a temporal and spatial order' within which there is 'a to-and-fro between temporality (succession, concatenation) and spatiality (simultaneity, synchronicity). This form is inseparable from orientation toward a goal ... the rationality of space' (*The Production of Space*, 71). Tosaka suggests a similarly dialectic approach, but not reduced in this way. As Tanabe had been influenced by Marxism, Tosaka seems to have taken quantum physics more seriously, following Tanabe's example, thus complicating matters of space and time.

16 Tosaka emphasizes the necessity of recognizing totalities, not universals.
17 TJz1, 380.
18 Tosaka argues that space itself must also be analyzed, and not only by analogy: 'space as analogy must not be confused with common sense conceptions of space' (TJz1, 494).
19 TJz1, 494.
20 TJz1, 374.
21 See Kōsaka Masaaki, *Japanese Thought in the Meiji Era* (Tokyo: Pan-Pacific Press, 1958), 84.
22 Tosaka, grounded in the science of various discourses on space, emphasizes what today has become the domain of Foucauvian studies: the modalities of separateness and unity in various types of discursive fields and arenas.
23 TJz1, 365, 367.
24 TJz2, 229.
25 TJz1, 371.
26 From 'About Space' (*Kūkan ni tsuite*, 1924). TJz1, 371.
27 TJz3, 39.
28 TJz3, 244.
29 TJz3, 245.
30 TJz3, 260.
31 TJz3, 261.
32 TJz3, 262 (emphasis in original).
33 TJz3, 263.
34 TJz1: 215ff. (italics reflect emphasis in original).
35 Here one sees Tosaka retrenching in a classical Marxist approach to materiality as primarily a social process based on historical materialist dialectics. Also note how his use of emphasis (underscoring assertion) may substitute for argument. Along the same lines, his use of long dashes to append sentences to sentences has no precedent in English and they are omitted, notably in this passage; however, other uses of long dashes in Tosaka are preserved where readability is not compromised (they are distinct elements of his performative writing style).
36 One must note that it is here, in making the link with 'cosmic time' (which suggests a return to a pre-relativity fixed space-time based on aether), that Tosaka's more-or-less orthodox Soviet situating of history is clear. The use of 'science' for such Marxists, and certainly in Tosaka's circle, subordinates the hard sciences to Marxian dialectical materialism. Moreover, one can indeed find a few purely scientific works by physicists (complete with formulae) that are devoid of historical context within the pages of the journal *Yuibutsuron kenkyū* (Studies in Materialism).
37 TJz3, 19.
38 TJz3, 19.
39 TJz3, 20.
40 TJz3, 22.
41 TJz3, 22–23.
42 TJz3: 25ff.
43 'Contrary' here may be translated as 'opposite' as well. Aristotelian logic includes many *opposites* ('Contraries (*enantia*), Contradictories (*apophaseis*), Possession and Privation (*hexis kai sterêsis*), Relatives (*pros ti*)'). See Robin

Smith, 'Aristotle's Logic', in Edward N. Zalta (ed.), *The Stanford Encyclopedia of Philosophy* (Summer 2019 Edition), https://plato.stanford.edu/archives/sum2019/entries/aristotle-logic/. However, here Tosaka uses Aristotelian *contraries* to situate physical movement within a Marxian dialectic. He mentions it only to turn to a more Kantian view of space and time as intuition, moving on to questions of how to situate the 'crisis' in physics.

44 There seems to be a conscious obfuscation in how 'movement' (*undō*) is used in this passage, in its sense of both motion in physics and development of ideas from one concept or paradigm to another in people's thoughts. Common sense is an important part of Tosaka's most famous work, *On Japanese Ideology* (1936). There he sees it as the centre of critical public space for vetting all sorts of competing ideas. In practice, however, Tosaka would adhere to the Soviet party line to an inordinate degree, influencing his understanding of contemporary discourse in physics (which remains highly contested in the philosophy of the foundations of physics to this day).

45 When the velocity of a physical body approaches the speed of light, according to Newtonian mechanics the relation of space and time are preserved (Tosaka's note).

46 Tosaka's wording here seems to suggest a peculiar confusion of the origins of relativity and its results, but I believe he is situating it in light of earlier thought on space that interests him and will be discussed later in this work.

47 When Tosaka situates the 'crisis' in modern physics this way, note how it naturalizes dialectics as part of a larger landscape of givens in the history of science and the world – as always grounded in facts – while at the same time it psychologizes the crisis, suggesting it is illusory. Compare Tanabe's discussion of the role of the subjective in quantum physics.

48 Tosaka is certainly identifying formal logic itself as part of a greater offence to materialism: idealism.

49 Though mathematization of course includes quantification (*ryōka*), scientific methods must dialectically incorporate formulae, and if those who assert that progress in science is to be attained by way of the *qualification* of scientific methods, they are without any doubt conceptual dialecticians. The realization of progress in physics seems to proceed without any connection to these sort of concepts. As for both physicists and scientists, these sort of 'dialecticians' pivoting away from the concept of method are entirely non-dialectical – due to what amounts to their completely ahistorical understanding (Tosaka's note).

50 Tosaka alludes to the Heisenberg uncertainty principle (see Box 3.18).

51 This sentence may be interpreted as Tosaka arguing that existence itself is an abstraction that in itself can be seen to support idealism, which here is deconstructed as in effect an habitual assumption that one can describe necessity and contingency in terms of causality (even after the advent of the new quantum physics).

52 Here Tosaka seems to be simply asserting his mantra of a dialectical panacea for all types of problems; nevertheless, his use of 'substantial' (*jisshitsuteki ni*) indicates a distinction with the nominal, so that he surely is trying to pull the distinction of necessity and contingency away from epistemological formulations and closer to ontological ones that are co-productive with

issues in space-time for which he has long expressed an interest (since his earliest published work).
53 For an analysis of the concept of laws of causality, the works of Engels and Lenin are valuable classics. See also László Rudas, *Mechnistische und dialektische Theorie der Kausalität* (*Unter dem Banner des Marxismus*, 1929) (Tosaka's note).
54 László Rudas critiques Bukharin's economics in light of various fundamental concepts in physics. See László Rudas, *Über einige Grundbegriffe der Mechanik und Dialektik. (Kritik der Bucharinschen Gesellschaftslehre)* (*Unter dem Banner des Marxismus*, 1930) (Tosaka's note). Nikolai Ivanovich Bukharin (1888–1938) was a leading Soviet revolutionary and theorist. In English, on the Hungarian communist and prominent editor, László Rudas (1885–1950), see his *Dialectical Materialism and Communism* (London: Labour Monthly, 1934).
55 The pun here works slightly better in Japanese, as *handōteki* (reactionary) suggests not only a political formation but causal physical relations and even psychological implications.
56 When Tosaka writes of an 'inner moment' one cannot help but note that he seems to be both setting up a straw man of inward consciousness as fallacy falling under the category of idealism and at the same time depending on Nishida's voluminous treatment of the moment from his earlier monograph on pure experience to his last works on the historical situation of multiple subjects in co-present moments of interaction that constitute the world(s). However, if Nishida zooms in on a more intertextually imbricated Hegelian dialectic of subject and object, Tosaka, by contrast, can be seen to blur the minutiae endemic to physics (as well as to Nishida) and situate dialectics as part of a bigger historical and social picture which is for him (and Marxists of his era who were sympathetic to Soviet revolutionary critiques of bourgeois class constructs) superior.
57 This sentence summarizes Tosaka's linking of science to history and ideology.
58 See Ema Michin, 'Lenin and the Crisis of the Natural Sciences in the Age of Imperialism', *Under the Banner of Marxism* 1, no. 2 (1931) (Tosaka's note, to a Japanese translation appearing in the journal). Also note that the illogical forcing of this argument may remind one of the excesses of poststructuralism, when a pun (here using 'contingent') echoes earlier discussions of 'contingency'.
59 Abram Moiseevich Deborin (1881–1963).

5 What We Can Learn from the Kyoto School

1 See John C. Maraldo, 'Nishida Kitarō', in Edward N. Zalta (ed.), *The Stanford Encyclopedia of Philosophy* (Winter 2019 Edition), https://plato.stanford.edu/archives/win2019/entries/nishida-kitaro/.
2 TJz1, 217.
3 Diana H. Coole and Samantha Frost, *New Materialisms: Ontology, Agency, and Politics* (Durham, NC, and London: Duke University Press, 2010), 10.

4 See, for instance, Valia Allori, 'Quantum mechanics and paradigm shifts', *Topoi* 34, no. 2 (2015): 313–23.
5 Kristian Camilleri, *Heisenberg and the Interpretation of Quantum Mechanics: The Physicist as Philosopher* (Cambridge: Cambridge University Press, 2009), 151.
6 Heisenberg, quoted in Camilleri, *Heisenberg and the Interpretation of Quantum Mechanics*, 148.
7 Niels Bohr, 'On the Notions of Causality and Complementarity', *Science*, New Series, 111, no. 2873 (20 January 1950): 52.
8 Bohr, 'On the Notions of Causality and Complementarity', 52.
9 Bohr, 'On the Notions of Causality and Complementarity', 53.
10 As will be shown below, this situation forms one aspect of a quantum dialectics discussed by Tanabe in the 1930s.
11 J. S. Bell, *Speakable and Unspeakable in Quantum Mechanics* (Cambridge: Cambridge University Press, 1987), 40–2, 174.
12 NKz7, 41. Translation by Gereon Kopf, 'Between Identity and Difference: Three Ways of Reading Nishida's Non-Dualism', *Japanese Journal of Religious Studies* 31, no. 1 (2004): 81–2.
13 James W. Heisig, *Philosophers of Nothingness: An Essay on the Kyoto School* (Honolulu: University of Hawai'i Press, 2001), 63 (italics in original).
14 Kopf, 'Between Identity and Difference', 80.
15 Kopf, 'Between Identity and Difference', 82.
16 Kopf, 'Between Identity and Difference', 92.
17 Georg W. Hegel and George di Giovanni, *The Science of Logic* (Cambridge: Cambridge University Press, 2010), 83 (italics in original).
18 Lucy Schultz, 'Nishida Kitarō, G. W. F. Hegel, and the Pursuit of the Concrete: A dialectic of Dialectics', *Philosophy East and West* 62, no. 3 (July 2012): 324.
19 Schultz, 'Nishida Kitarō, G. W. F. Hegel, and the Pursuit of the Concrete', 319.
20 Georg W. Hegel and M. J. Petry, *Philosophy of Nature*, vol. 2 (London: George Allen and Unwin, 1970), 229 (§260) (bold and italics in original).
21 Kitarō Nishida, *Last Writings: Nothingness and the Religious Worldview*, trans. David A. Dilworth (Honolulu: University of Hawaii Press, 1987), 52; NKz11, 377.
22 Nishida, *Last Writings*, trans. Dilworth (slightly altered), 92; NKz11, 427.
23 NKz4, 241.
24 NKz4, 244.
25 NKz4, 245.
26 Krummel, *Nishida Kitarō's Chiasmatic Chorology*, 200.
27 See Chapter 2.
28 THz14, 462.
29 THz14, 464.
30 THz14, 465.
31 THz14, 459. Indeed, the relation of classical and quantum physics has yet to sort out Wilfrid Sellars' distinction between '*scientific image* (the image of the world that our best scientific theories are giving us)' and 'the *manifest image* (the image of the world that we ordinarily experience)'. Valia Allori, *The Wave Function: Essays on the Metaphysics of Quantum Mechanics*, ed. Alyssa Ney and David Z. Albert (Oxford: Oxford University Press, 2013), 59 (italics in original).
32 THz14, 459.

33 Moreover, Tanabe's approach to dialectics in light of the new physics' interest in an *ontological* foundation that is not merely objective or descriptive, since, 'there is absolutely no basis for establishing a dialectics out of either existence or recognition' if building on 'an image passively described' (THz14, 466). It is 'only by way of nature and recognition working together that dialectical unity begins to form' (466).
34 THz5, 323.
35 THz5, 322.
36 THz5, 439–40.
37 THz5, 441.
38 Barad, *Meeting the Universe Halfway*, 89.
39 Daniela Frauchiger and Renato Renner, 'Quantum theory cannot consistently describe the use of itself', *Nature Communications* 9, no. 3711 (2018): 1.
40 See Allori, 'Quantum mechanics and paradigm shifts', 313–23; Tim Maudlin, *Philosophy of Physics: Quantum Theory* (Princeton, NJ: Princeton University Press, 2019); and Reza Maleeh and Parisa Amani, 'Pragmatism, Bohr, and the Copenhagen Interpretation of Quantum Mechanics', *International Studies in the Philosophy of Science* 27, no. 4 (2013): 353–67.
41 Schrödinger's cat has become within physics a sort of stale folk tale that may have outlived its pedagogical usefulness, since it provides a negative example of how elaborate examples distract from understanding problems of superpositioning – states that are difficult to observe simultaneously and as such should themselves be elaborated upon. This thought experiment featuring a living or dead cat in an unopened box is not half alive or dead, as it is a macrological object in a causally-triggered experimental set-up beyond the ken of the quanta. Its core cruelty to animals distracts from its core position of simply adhering to classical physical expectations.
42 Frauchiger and Renner, 'Quantum theory cannot consistently describe the use of itself', 2.
43 Maleeh and Amani, 'Pragmatism, Bohr, and the Copenhagen Interpretation of Quantum Mechanics', 355, paraphrasing Murdoch, *Niels Bohr's Philosophy of Physics*, 105. See also Alexander Wendt, *Quantum Mind and Social Science: Unifying Physical and Social Ontology* (Cambridge: Cambridge University Press, 2015), 74.
44 Slobodan Perovic, 'Why were matrix mechanics and wave mechanics considered equivalent?', *Studies in History and Philosophy of Modern Physics* 39 (2008): 445.
45 Wendt, *Quantum Mind and Social Science*, 238.
46 Wendt, *Quantum Mind and Social Science*, 239.
47 Camilleri, *Heisenberg and the Interpretation of Quantum Mechanics*, 142ff.
48 Ulf Mellström, 'From a Hegemonic Politics of Masculinity to an Ontological Politics of Intimacy and Vulnerability? Ways of Imagining through Karen Barad's Work', *Rhizomes: Cultural Studies in Emerging Knowledge* 30 (2016): para. 5.
49 Barad, *Meeting the Universe Halfway*.

Timeline of Kyoto School Activities in Relation to Physics

Milestones in physics

1887 Michelson–Morley experiment disproves assumption of aether as a universal medium.
1887 Heinrich Rudolf Hertz conjoins electricity and magnetism as electromagnetic waves.
1889, 1892 Length (or Lorentz-FitzGerald) contraction: an ad hoc theory to explain the absence of aether.

1896 Henri Becquerel discovers radioactivity.
1897 J. J. Thomson discovers the electron.

1900 Max Planck works from assumptions of black-body radiation to develop origins of quantifying light, initiating quantum physics.
1905 Albert Einstein develops special relativity, proposes wave-particle duality in light and idea of the photon to explain the photoelectric effect and Brownian motion.

Japan and the Kyoto School
– Nishida, Tanabe and Tosaka –

1868 Reformist intellectual Fukuzawa Yukichi (1835–1901) publishes *Illustrated Book of Physical Sciences* and many other works introducing Western cultures.
1870 Nishida Kitarō born 19 May, Ishikawa Prefecture.
1885 Tanabe Hajime born 3 February, Tokyo.
1887 Nishida chooses German as second foreign language (English used in advanced classes in school).

1896 Nishida begins to study Zen Buddhism.
1888 Nishida wavers between majoring in maths and philosophy for college-level studies (called 'high school' under the old system), ultimately choosing the latter.
1899 Nishida teaches psychology, ethics, logic and German as instructor.

1900 Tosaka Jun born 7 September, Tokyo.
1906 Nishida begins to publish essays that would appear in his first book.
1908 Tanabe enrols in Tokyo Imperial University mathematics department, later transfers to the philosophy department.
1910 Nishida appointed assistant professor at Kyoto Imperial University.

Milestones in physics	Japan and the Kyoto School – Nishida, Tanabe and Tosaka –
	1911 Nishida publishes *A Study of Good* (*Zen no kenkyū*).
	1912 Tanabe publishes 'On Relativity' and other articles related to the philosophy of science in *Tetsugaku zasshi*.
1911 Ernest Rutherford discovers the atomic nucleus.	1913 Nishida publishes 'History and the Natural Sciences' in *Tetsugaku zasshi*.
1911 Albert Einstein: Equivalence Principle.	1913–17 Nishida publishes 'Intuition and Reflection in Self-consciousness' (*Jikaku ni okeru chokkan to hansei*), serialized in *Geibun* and *Tetsugaku kenkyū*.
1913 Niels Bohr: Bohr model of the atom.	
	1915 Nishida publishes *Thinking and Experience* (*Shisaku to taiken*).
	1915 Tanabe publishes *The Contemporary Natural Sciences* (*Saikin no shizen kagaku*).
	1915 Tanabe publishes translation of Poincaré's 1905 *The Value of the Sciences* (*Kagaku no kachi*).
	1917 Nishida publishes *Intuition and Reflection in Self-consciousness* (*Jikaku ni okeru chokkan to hansei*) as a book.
1916 Albert Einstein outlines his general theory of relativity.	1917 Tanabe publishes 'On Time' in *Tetsugaku zasshi*.
1919 The bending of light confirms evidence for general relativity.	1918 Tanabe publishes *Outline of the Sciences* (*Kagaku gairon*).
	1918 Tosaka aspires to be a physicist and, upon entering college-level studies, majors in mathematics in the science department and develops an interest in fundamental problems in the natural sciences.
	1919 Nishida invites Tanabe to become an assistant professor at Kyoto Imperial University.

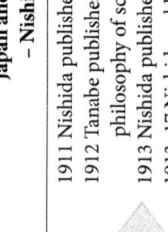

1922 Stern–Gerlach experiment helps define quantifiable spatial orientation of angular momentum (spin) in elementary particles.

1924 Louis de Broglie postulates matter waves (or de Broglie waves), thus contributing to the ongoing refinement of Einstein's observation of wave–particle duality.

1923 Arthur Compton discovers the particle nature of photons and the Compton effect.

1924 Bose–Einstein statistics

1925 Werner Heisenberg conceives of matrix mechanics to situate particles in quantum mechanical systems.

1926 Erwin Schrödinger formulates Schrödinger Equation to situate wave functions in quantum mechanical systems.

1927 Werner Heisenberg postulates the uncertainty principle (see Box 3.18).

1927 Dirac equation integrates relativity theory in quantum determination processes.

1928 Paul Dirac proposes the antiparticle.

1920 Nishida publishes *The Problem of Consciousness* (*Ishiki no Mondai*).

1921 Idolizing Nishida and Tanabe, after graduation, Tosaka enrols in the philosophy department at Kyoto Imperial University, majoring in mathematic philosophy, and continuing his study of the foundations of discourse on space in the natural sciences.

1922 Einstein visits Japan at Nishida's recommendation and gives his first talk at Kyoto Imperial University on the origins of relativity.

1922 Tosaka pens 'The Realization of Physical Space'.

1924 Nishida publishes 'What Lies Behind Physical Phenomena' in *Shisō* (see full translation in 2.1).

1924 Tosaka graduates from university and continues with graduate studies.

1924 Tosaka publishes 'Kant and Contemporary Science' and 'Kant's Theory of Space' in *Tetsugaku kenkyū*.

1925 Tosaka publishes 'The Realization of Physical Space' in *Tetsugaku kenkyū*.

1925 Peace Preservation Law dampens period of flourishing democratic thought by limiting freedom of speech and assembly.

1926 Tosaka publishes 'Geometry and Space' in *Shisō*, and 'Concerning Space as a Category' in *Tetsugaku kenkyū*.

1927 Nishida publishes *From the Acting to the Seeing* (*Hatarakumono kara mirumono e*), which includes a chapter translated in 2.1 above.

1929 Tosaka becomes a professor at Otani University and a lecturer at Kobe University of Commerce (part of Kobe University today).

1929 Tosaka publishes *On the Scientific Method* and articles including 'The Historical Social Constraints on Science' in *Tōyō gakugei zasshi*.

Milestones in physics	Japan and the Kyoto School – Nishida, Tanabe and Tosaka –
	1927 Tanabe 'The Logic of Dialectics' serialized in *Tetsugaku zasshi*.
	1930 Nishida publishes *The Self-aware System of Universals* (*Ippansha no jikakuteki taikei*).
	1930 Tanabe publishes famous criticism of Nishida, 'Looking Up to the Teachings of Nishida', in *Tetsugaku zasshi*, marking the turn toward development of Tanabe's own thought and the Kyoto School itself.
	1930 Tosaka publishes 'The Popularity of Science' and 'Nature Dialectics' in *Shisō*.
	1931 Tosaka publishes many works on Kant and Hegel's work on the philosophy of science and more writings on space. He also publishes many articles focusing on ideology as well as philosophy in the Soviet Union.
	1932 Nishida publishes *The Self-Aware Determination of Nothingness* (*Mu no jikakuteki gentei*).
1932 Carl David Anderson discovers antimatter.	1932 Tosaka publishes 'The Formation of Tanabe Philosophy' in *Shisō*.
1932 James Chadwick discovers neutron.	1932–8 *Yuibutsuron kenkyū* (*Studies in Materialism*), reflecting the work of members of the Tosaka's Materialist Research Association (*Yuibutsuron kenkyū-kai*) and affiliates, publishes under the leadership of Tosaka, with philosophy of science articles mostly embracing Communist Party positions that refuse to accept the findings of quantum mechanics (with the exception of a few articles of a technical nature by physicists).
	1933 Nishida publishes *Fundamental Problems of Philosophy: The World of Action* (*Tetsugaku no konpon mondai: kōi no sekai*).
	1933 Tanabe publishes 'The New Physics and Nature Dialectics' (see full translation in 3.1).
	1933 Tosaka publishes a famous critique of Nishida, 'Is *Logic of Nothingness* a Logic', in *Yuibutsuron kenkyū*.
	1934 Nishida publishes *Fundamental Problems of Philosophy Continued: The Dialectical World* (*Tetsugaku-no konpon mondai zokuhen: Benmsyouhouteki-sekai*).

1935 Hideki Yukawa (1907–81) postulates a theory of mesons (elementary particles) for which he was awarded a Nobel Prize in Physics. Professor at Kyoto University (1940–9).
1937 Anderson and Neddermeyer discover muon.
1938 Nuclear fission discovered (leading to production of atomic bombs).

1942–6 The Manhattan Project developed atomic weapons under the leadership of nuclear physicist Robert Oppenheimer.

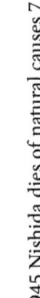

1935 Tosaka publishes *On Japanese Ideology* and *On Science*.
1935–46 Nishida's philosophical writings in this period are collected in volumes (1–7) titled *Collected Philosophical Writings* (*Tetsugaku-teki ronbun-shū*).
1936 Tanabe publishes 'The Philosophical Significance of Quantum Theory' in *Shisō*, and a major book, *Between Philosophy and Science* (*Tetsugaku to kagaku to no aida*; see 3.2 above for excerpts).
1935–7 Purge of leftist academics and officials.
1937 Tosaka publishes 'On Contemporary Science Education' in *Kagaku pen*.
1938 Tosaka arrested for violating the Peace Preservation Law.
1939 Nishida first publishes 'The Empirical Sciences' in *Shisō* (see excerpts in 2.2).

1944 Nishida publishes 'The World of Physics' in *Shisō*.

1945 Nishida dies of natural causes 7 June.
1945 US aircraft explode atomic bombs over Hiroshima on 6 August and over Nagasaki on 9 August.
1945 Tosaka dies due to intense heat in his prison cell on 9 August.

Glossary

Absolute contradictory self-identity *zettai mujunteki jiko-dōitsu* 絶対矛盾的自己同一

Absolute nothingness *or* **absolute nothing** *zettai mu* 絶対無

Absolute self *or* **absolute ego** *zettai-ga* 絶対我 (in Fichte, but echoing Buddhism, the possibility of self and non-self, before a self is situated in the world through action; rather than an indelible self)

Act; action; activity; influence; operation; process; function; agency *sayō* 作用

Act of observation *or* **measurement** *kansoku-kōi* 観測行為

Active (1) *kōiteki* 行為的 (Kyoto School uses of this term convey a sense of engaging in action rather than degree of intensity of as action); (2) *nōdō* 能働的 (used in contrast with passivity)

Active/dynamic self *kōdōteki jiko* 行動的自己

Active impressions *nōdōteki inshō* 能働的印象

Active intuition *kōiteki chokkan* 行為的直感; also, actively intuitive *kōiteki chokkanteki*

Alternation *kōgosei* 交互性 (may also be translated as mutuality or reciprocity, but Tanabe uses it to approach the concept of uncertainty in the uncertainty principle, and later complementarity)

Appearance of truth *shinsō* 真相

Atomic; indivisibility *fukabun* 不可分

Attribute *zokusei* 属性

Basho *basho* 場所 (may be translated variously as place, logos, matrix or field of immanent relations)

(A) bound up with (B) *soku* 即 (negation joined with affirmation between two items; translation controversial, sometime rendered *qua*; here also translated as '(A) joined with (B)'; 'the inseparability of (A and B)'; 'both (A and B) indivisibly')

Cognition *ninshiki* 認識

Commutation point *tenkanten* 転換点

Complementarity *sōhosei* 相補性 (Bohr's influential terms similar to Heisenberg's uncertainty principle, but stated as a more ontologically significant binding of canonical conjugate pairs of observables (such as position and momentum) within a given quantum phenomenon that cannot be approached simultaneously due to limitations encountered in the processes of observation and demanding a splitting of measurement times according to diverse laboratory set-ups)

Conduct *kōi* 行為

Configure *kōsei suru* 構成 する

Consciousness *ishiki* 意識

Contents *naiyō* 内容 (usually translated as 'contents', yet sometimes may more idiomatically be rendered 'significance'.)
Contradictory self-identity *mujunteki jiko dōitsu* 矛盾的自己同一
Copenhagen Interpretation The descriptive historical name given to approaches to quantum physics deriving in the work of Bohr and Heisenberg, especially complementarity and the uncertainty principle.
Determination *gentei* 限定
Dialectics of nature *shizen benshōhō* 自然弁証法
Distinction, identifying distinction *shikibetsu* 識別
Elements of experimental knowledge *jikkenjō sokuchi no yōso* 実験上即知の要素
Event *dekigoto* 出来事
Expressive agential *hyōgen sayōteki* 表現作用的 (lit. expression-agential)
Expressive universal *hyōgenteki ippansha* 表現的一般者
Field *ba* 場 (*ba* in contexts outside of physics also has meanings including 'place' and 'situation')
Here and now *ima/koko* 今・此処
Hole (of electron); positive hole *seikō* 正孔; also *kūkō* 空孔 (used in Dirac's theory of the hole that led to the discovery of the positron; later known simply as the positron)
Ideal causality *rinen inga* 理念因果
Immediately; directly *chokusetsuteki ni* 直接的に
In a self-contradictory way *jikohiteiteki ni* 自己否定的に
Incorporate; contain within *fukumu* 含む
Individual (n.) depending on the context, it may be used to translate:
(1) *kotai* 個体 *individuum* (distinct from others; indivisible; existing independently; individual thing/body)
(2) *kojin* 個人 individual person (always refers to an individual person)
Individual (things) *kobutsu* 個物 (particular(s))
Indivisibility; being bonded; bound to one another *sōsoku* 相即, dissolving into oneness (as in Buddhism); being inseparable (as in complementarity in quantum physics)
Inseparable; strongly attached to each other *sōsokufuri* 相即不離 (also see entry for 'bound up with' (*soku*))
Interpenetration and indivisibility; mutually bound *sōnyū-sōsoku* 相入相即 (related to Buddhist term *sōsoku-sōnyū*)
Introspection *naisei* 内省
Local, locales *koko* 此処 (this keyword might be translated as 'here' or 'this place' or even simply 'a place'; however, to avoid confusion with Nishida's use of *basho* (often translated as place), local and locale are preferred in most renderings)
Matter *shitsuryō* 質料 (in an Aristotelian sense of hylomorphism, which sees matter and form as joined in being)
Matter wave *busshitsu-ha* 物質波
Mental image *shinzō* 心像
Mental phenomenon *shinri genshō* 心理現象

Moment *keiki* 契機 (can also mean *opportunity*); *junkan* 瞬間 (also *instant*)
Nature in physics; nature in terms of physics; physical nature *butsurigakuteki shizen* 物理学的自然
Nothingness (1) *mu* (2) *kyomu* 虚無 (with nihilistic sense of void)
Object world *taishō-kai* 対象界
Objective world *kyakkanteki(na)sekai* 客観的(な)世界
Oblique viewing mode *shashi yōsō* 斜視様相
Operation(al) *sōsa(teki)* 操作(的)
Perception *chikaku* 知覚
Phenomenon of consciousness *ishiki genshō* 意識現象
Physical force *butsuryoku* 物力
Position *ichi* 位置
Predicate *jutsugo* 述語 (in propositional logic in contrast to the subject)
Present *ima* 今
Present (v.); presentation *hyōshō suru* 表象する; *hyōshō* 表象 (in sense of the German word *Vorstellung*)
Realm of objectivity *kyakkan-kai* 客観界
Render *mosha suru* 模写する; lit. copy, reproduce
Self-awareness; consciousness of will *ishi no jikaku* 意志の自覚
Self-awareness of the acting *hataraku mono no jikaku* 働くものの自覚
Self-consciousness *jikaku no ishiki* 自覚の意識
Self-consciousness awareness *jikaku* 自覚
Selflessness *muga* 無我 (one of the three or four most important concepts in Buddhism (along with impermanence, suffering and nirvana), translated from the Sanskrit *Anatta*, the not-self or the seeing beyond the attachments of the fiction of a self)
Self-reflection *shisei* 自省
Sensation *kankaku* 感覚
Sensory *kankaku-teki* 感覚的
Significance *imi* 意味 or *igi* 意義 (usually translated as 'meaning' and 'significance' respectively)
Space-time *jikū* 時空 or *kūkanjikan* 空間時間
Standard measure *shakudo* 尺度
Standpoint of a state of oneness *dōsei ichi'nyo no tachiba* 動静一如の立場 (Note: *ichi'nyo* has Buddhist origins)
Subject (1) *shutai* 主体 (the situated human subject)
(2) *shugo* 主語 (in propositional logic in contrast to the predicate)
Superpositioning The simultaneous existence of observables associated with a given (repeatable) phenomenon that nevertheless cannot be reduced to one phenomenal measurement due to the role of the observing means and apparatus in determining them.
Supra-conscious world *chō-ishiki-kai* 超意識界
Supra-individual *chō-kojinteki* 超個人的
Uncertainty principle *fukakuteisei genri* 不確定性原理 (also known as the indeterminacy principle) (see Box 3.18)

Understanding of nature *shizen ninshiki* 自然認識 (also, 'understanding the natural environment')
Unhinged *dōyō* 動揺 (lit. disturbed; in commotion; discomposed)
Viewing mode *chokushi yōsō* 直視様相
Visual sensory activity *shikaku sayō* 視覚作用
Void *kū* 空
Yin and yang *inyō* 陰陽 (ancient Chinese form of dualism for situating universal phenomena according to complementary and interdependent pairs, such as light (yang) and dark (yin); a standard for making distinctions)

Further Reading

Online resources for science and philosophy (general)

Chronology of Milestone Events in Particle Physics (as it reflected in scientific literature). Maintained by COMPAS (Institute for High Energy Physics) and Particle Data Group (PDG) at Lawrence Berkeley National Laboratory. URL = http://web.ihep.su/dbserv/compas/contents.html. A useful site for clearly pinpointing how and when particular discoveries have been documented within the field of physics and the philosophy of physics. Includes links to cited documents going back over 120 years.

General Philosophy of Science, Howard Sankey (ed.) (University of Melbourne). URL = https://philpapers.org/browse/general-philosophy-of-science. The leading public repository of philosophical papers, focused on this vast subtopic.

Philosophy of Science. URL = https://www.journals.uchicago.edu/toc/phos/current. Journal sponsored by the Philosophy of Science Association.

Physics Stack Exchange. URL = https://physics.stackexchange.com/. A user-driven website for the exchange of questions and answers related to physics.

Quanta Magazine. URL = https://www.quantamagazine.org/. A journal devoted to the public understanding of the latest developments in scientific discovery and applications in sophisticated language with minimal technical equations and many illustrations.

Science Daily. URL = https://www.sciencedaily.com/. A comprehensive website collecting primarily public relations releases by prominent laboratories around the world, including links to research articles.

The Stanford Encyclopedia of Philosophy, Edward N. Zalta (ed.). URL = https://plato.stanford.edu. A rich source of regularly updated articles on key thinkers and issues in philosophy in general; in a broad sense inclusive of scientific thought and some world philosophies.

Works relating to the Kyoto School and Japanese philosophy in general

Carter, Robert E., *The Kyoto School: An Introduction*. Albany, NY: State University of New York Press, 2013.

Davis, Bret W., 'The Kyoto School', in Edward N. Zalta (ed.), *The Stanford Encyclopedia of Philosophy* (Summer 2019 edition), https://plato.stanford.edu/archives/sum2019/entries/kyoto-school/. A helpful compressed overview of the Kyoto School, including an extensive bibliography of translations and secondary literature.

Davis, Bret W., *The Oxford Handbook of Japanese Philosophy*, New York: Oxford University Press, 2020.

Davis, Bret W., Brian Schroeder and Jason M. Wirth, *Japanese and Continental Philosophy: Conversations with the Kyoto School*, Bloomington: Indiana University Press, 2011.
Dilworth, David A., V. H. Viglielmo, and Agustín Zavala, *Sourcebook for Modern Japanese Philosophy: Selected Documents*, Westport, Conn: Greenwood Press, 1998.
Doak, Kevin M., 'Under the Banner of the New Science: History, Science, and the Problem of Particularity in Early Twentieth-Century Japan', *Philosophy East and West* 48, no. 2 (April 1998): 232–56.
Fujita, Masakatsu, Robert Chapeskie and John W. Krummel, *The Philosophy of the Kyoto School*, Gateway East, Singapore: Springer, 2018.
Heisig, James W., *Much Ado about Nothingness: Essays on Nishida and Tanabe*, Nagoya, Japan: Nanzan Institute for Religion and Culture, 2015.
Heisig, James W., *Philosophers of Nothingness: An Essay on the Kyoto School*, Honolulu: University of Hawai'i Press, 2001.
Heisig, James W. and John C. Maraldo, *Rude Awakenings: Zen, the Kyoto School, and the Question of Nationalism*, Honolulu: University of Hawai'i Press, 1995.
Itabashi, Yūjin, 'Grounded on Nothing: The Spirit of Radical Criticism in Nishida's Philosophy', *Philosophy East and West* 68, no. 1 (2018): 97–111.
Jones, Christopher S., *Re-politicising the Kyoto School as Philosophy*, London and New York: Routledge, 2008.
Kasulis, Thomas P., 'Sushi, Science, and Spirituality: Modern Japanese Philosophy and Its Views of Western Science', *Philosophy East and West* 45, no. 2 (April 1995): 227–48.
Krummel, John W., 'Chōra in Heidegger and Nishida', *Studia Phænomenologica* 16 (2016): 489–518.
Krummel, John W., *Nishida Kitarō's Chiasmatic Chorology: Place of Dialectic, Dialectic of Place*, Bloomington: Indiana University Press, 2015.
Krummel, John W., *Contemporary Japanese Philosophy: A Reader*, Lanham, Maryland: Rowman & Littlefield, 2019.
Maraldo, John C., *Japanese Philosophy in the Making 1: Crossing Paths with Nishida*, Studies in Japanese Philosophy, ed. Takeshi Morisato, Nagoya, Japan: Chisokudō, 2017.
Maraldo, John C., *Japanese Philosophy in the Making 2: Borderline Interrogations*. Nagoya: Chisokudō Publications, 2019.
Maraldo, John C., 'Nishida Kitarō', in Edward N. Zalta (ed.) *The Stanford Encyclopedia of Philosophy* (Winter 2019 Edition), https://plato.stanford.edu/archives/win2019/entries/nishida-kitaro/.
Maraldo, John C., 'The War over the Kyoto School', *Monumenta Nipponica* 61, no. 3 (2006): 375–406.
Suares, Peter. *The Kyoto School's Takeover of Hegel: Nishida, Nishitani, and Tanabe Remake the Philosophy of Spirit*, Lanham, MD: Lexington Books, 2011.
Takeshi, Morisato, *Critical Perspectives on Japanese Philosophy*, Nagoya, Japan: Nanzan, 2016.
Waldenfels, Hans, 'Absolute Nothingness: Preliminary Considerations on a Central Notion in the Philosophy of Nishida Kitarō and the Kyoto School', *Monumenta Nipponica* 21, no. 3/4 (1966): 354–91, doi:10.2307/2383379. Dated, but raises important issues between Nishida and Tanabe and their positions.

Williams, David, *Defending Japan's Pacific War: The Kyoto School Philosophers and Post-White Power*, London and New York: Routledge Curzon, 2004.
Yusa, Michiko, *The Bloomsbury Research Handbook of Contemporary Japanese Philosophy*, London: Bloomsbury Academic, 2017.

Works relating to Nishida Kitarō

Arisaka, Yoko, 'Beyond "East and West": Nishida's Universalism and Postcolonial Critique', *Review of Politics* 59, no. 3 (1997): 541–60.
Arisaka, Yoko and Nishida Kitaro, 'The Nishida Enigma: "The Principle of the New World Order"', *Monumenta Nipponica* 51, no. 1 (1996): 81–105.
Cestari, Matteo, 'Between Emptiness and Absolute Nothingness: Reflections on Negation in Nishida and Buddhism', in James W. Heisig and Rein Raud (eds), *Frontiers of Japanese Philosophy: Japanese Philosophy Abroad*, 320–46, Nagoya: Nanzan Institute for Religion and Culture, 2010.
Cestari, Matteo, 'The Knowing Body: Nishida's Philosophy of Active Intuition (Kōiteki Chokkan)', *Eastern Buddhist*, New Series, 31, no. 2 (1998): 179–208.
Dilworth, David, 'The Initial Formations of "Pure Experience" in Nishida Kitarō and William James', *Monumenta Nipponica* 24, no. 1/2 (1969): 93–111, doi:10.2307/2383764.
Krueger, Joel W., 'The Varieties of Pure Experience: William James and Kitaro Nishida on Consciousness and Embodiment', *William James Studies* 1, https://williamjamesstudies.org/the-varieties-of-pure-experience-william-james-and-kitaro-nishida-on-consciousness-and-embodiment/.
Maraldo, John C., 'Nishida Kitarō', in Edward N. Zalta (ed.), *The Stanford Encyclopedia of Philosophy* (Winter 2019 Edition), https://plato.stanford.edu/archives/win2019/entries/nishida-kitaro/. Here is both an excellent introduction to Nishida Philosophy and helpful bibliographies of translations of Nishida into various European languages and of important secondary literature.
Nishida, Kitarō, *An Inquiry into the Good*, New Haven, CT: Yale University Press, 1990.
Nishida, Kitarō, *Last Writings: Nothingness and the Religious Worldview*, trans. David A. Dilworth, Honolulu: University of Hawaii Press, 1987.
Nishida, Kitarō, *Nishida Kitarō zenshū*, 19 vols, Tokyo: Iwanami Shoten, 1989. Abbr. NKz.
Nishida, Kitarō, *Place and Dialectic: Two Essays by Nishida Kitarō*, trans. John W. M. Krummel and Shigenori Nagatomo, Oxford: Oxford University Press, 2012.
Nishida, Kitarō and William W. Haver, *Ontology of Production: Three Essays*, Durham, NC: Duke University Press, 2012.
Raud, Rein, '"Place" and "Being-Time": Spatiotemporal Concepts in the Thought of Nishida Kitaro and Dogen Kigen', *Philosophy East and West* 54, no. 1 (2004): 29–51.
Rigsby, Curtis A., 'Nishida on Heidegger', *Continental Philosophy Review* 42 (2010): 511–53.
Schultz, Lucy, 'Nishida Kitarō, G. W. F. Hegel, and the Pursuit of the Concrete: A dialectic of Dialectics', *Philosophy East and West* 62, no. 3 (July 2012): 319–38.

Scribner, F. Scott., 'Nishida's Fichte and the Resistance of Idealism', *International Journal for Field-Being* 1, no. 1, Part 2, Article No. 13 (2001), www.iifb.org/ijfb/FSSScribner-2-13.
Tani, Toru, 'Inquiry into the I, disclosedness, and self-consciousness: Husserl, Heidegger, Nishida', *Continental Philosophy Review* 31 (1998): 239–53.
Tremblay, Jacynthe, *Je suis un lieu*, Montréal: Les Presses de l'Université de Montréal, 2016.
Tremblay, Jacynthe, 'L'influence du concept de complémentarité dans la philosophie du dernier Nishida', *European Journal of Japanese Philosophy* 3 (2018): 57–77.
Wargo, Robert J. J., *The Logic of Nothingness: A Study Of Nishida Kitarō*, Honolulu: University of Hawai'i Press, 2005.
Yusa, Michiko, *Zen and Philosophy: An Intellectual Biography of Nishida Kitaro*, Honolulu: University of Hawai'i Press, 2002. Includes discussion of impact of Einstein as well as Zen on Nishida.

Works related to authors cited by Nishida Kitarō

Bergson, Henri, *Time and Free Will: An Essay on the Immediate Data of Consciousness*, London: George Allen and Unwin, 1910.
Bergson, Henri, *Creative Evolution*, New York: Dover, 1998 (1911).
Brentano, Franz, Stephan Körner, and Roderick M. Chisholm, *Philosophical Investigations on Space-time and the Continuum*, London: Routledge, 2010.
Brentano, Franz, Oskar Kraus and Linda L. McAlister, *Psychology from an Empirical Standpoint*, London and New York: Routledge, 1995.
James, William, *Essays in Radical Empiricism*, Lincoln: University of Nebraska Press, 1996 (1912).
James, William, *A Pluralistic Universe*. Cambridge, MA: Harvard University Press, 1977 (1909).
James, William, *Psychology: The Briefer Course*, Mineola, NY: Dover Publications, 2001 (1892).
Leonard, Lawlor and Valentine Moulard Leonard, 'Henri Bergson', in Edward N. Zalta (ed.), *The Stanford Encyclopedia of Philosophy* (Summer 2016 edition), https://plato.stanford.edu/archives/sum2016/entries/bergson/.
Lewin, Kurt, *A Dynamic Theory of Personality*, trans. Donald K. Adams and Karl E. Zener, New York and London: McGraw-Hill Book Company, 1935.
Lewin, Kurt, *Principles of Topological Psychology*, trans. Fritz Heider and Grace M. Heider, New York and London: McGraw-Hill Book Company, 1936.
Truman, Nathan Elbert, *Maine de Biran's Philosophy of Will*, New York: Macmillan Company, 1904.

Works relating to Tanabe Hajime

Heisig, James W, 'Tanabe's Logic of the Specific and the Critique of the Global Village', *Eastern Buddhist*, New Series, 28, no. 2 (1995): 198–224. See also Heisig's books on the Kyoto School.

Morisato, Takeshi, 'Translation of Tanabe Hajime's "The Limit of Logicism in Epistemology: A Critique of the Marburg and Freiburg Schools"', *Journal of World Philosophies* 2 (2017): 1–26.

Osaki, Makoto, *Individuum, Society, Humankind: The Triadic Logic of Species according to Hajime Tanabe*, Leiden: Brill, 2001.

Sakai, Naoki, 'Subject and Substratum: On Japanese Imperial Nationalism', *Cultural Studies* 14, no. 3/4 (2000): 462–530. An in-depth exploration of Tanabe in relation to political thought in English.

Tanabe, Hajime, *Tanabe Hajime zenshū*, 15 vols, Tokyo: Chikuma Shobō, 1964. Abbr. THz.

Tanabe, Hajime, *Philosophy as Metanoetics*, trans. Takeuchi Yoshinori, Valdo Viglielmo and James W. Heisig, University of California Press, 1987.

Tanabe, Hajime, David Dilworth and Taira Satō, 'The Logic of the Species as Dialectics', *Monumenta Nipponica* 24, no. 3 (1969): 273–88.

Works relating to Tosaka Jun

Tosaka, Jun (戸坂 潤), *Tosaka Jun zenshū* (The complete works of Tosaka Jun), 5 vols, Tokyo: Keisōshobō (勁草書房), 1966–7. Abbreviated TJz.

Tosaka, Jun et al. *Tosaka Jun: A Critical Reader*, Ithaca, NY: East Asia Program, Cornell University, 2013. While avoiding problems of modern physics, this work introduces other central concepts, including the role of everyday space.

Works relating to issues in modern physics and the philosophy of science

Barad, Karen, *Meeting the Universe Halfway: Quantum Physics and the Entanglement of Matter and Meaning*, Durham, NC: Duke University Press, 2007.

Bell, J. S., *Speakable and Unspeakable in Quantum Mechanics: Collected Papers on Quantum Philosophy*, Cambridge: Cambridge University Press, 1987.

Beller, Mara, *Quantum Dialogue: The Making of a Revolution*, Chicago: University of Chicago Press, 1999.

Bitsakis, Eftichios I., 'Complementarity: Dialectics or Formal Logic?' *Nature, Society, and Thought* 15, no. 3 (2002): 275–306.

Bohm, David and F. David Peat, *Science, Order and Creativity*, London: Routledge, 2010.

Bohr, Niels et al., *Atomic Theory and the Description of Nature*, reprint, Cambridge: Cambridge University Press, 1961.

Bohr, Niels et al., *Collected works Volume 6: Foundations of Quantum Physics I (1926–1932)*, Amsterdam: North-Holland Publishing Company, 1972.

Bohr, Niels et al., 'On the Notions of Causality and Complementarity', *Science*, New Series, 111, no. 2873 (20 January 1950): 51–4.

Bokulich, Alisa, 'Paul Dirac and the Einstein-Bohr Debate', *Perspectives on Science* 16, no. 1 (2008): 103–14.

Bridgman, P. W., *The Logic of Modern Physics*, New York: Macmillan, 1927.

Camilleri, Kristian, *Heisenberg and the Interpretation of Quantum Mechanics: The Physicist as Philosopher*, Cambridge: Cambridge University Press, 2009.
Carroll, Sean, *The Particle at the End of the Universe: How the Hunt for the Higgs Boson Leads Us to the Edge of a New World*, New York: Dutton, 2012.
Dirac, P. A. M., *The Principles of Quantum Mechanics*, 4th ed. (rev.), Oxford: Clarendon Press, 1958.
Emery, A., 'Dialectics versus Mechanics: A Communist Debate on Scientific Method', *Philosophy of Science* 2, no. 1 (1935): 9–38.
Esfeld, Michael and Nicholas Gisin, 'The GRW Flash Theory: A Relativistic Quantum Ontology of Matter in Space-time?', *Philosophy of Science* 81 (2014): 248–64.
Faye, Jan and Henry J. Folse (eds), *Niels Bohr and Contemporary Philosophy*, Dordrecht: Kluwer Academic Publishers, 1994.
Faye, Jan and Henry J. Folse (eds), *Niels Bohr and the Philosophy of Physics: Twenty-first-century Perspectives*, London: Bloomsbury Academic, 2017.
Folse, Henry J., *The Philosophy of Niels Bohr: The Framework of Complementarity*. Amsterdam and New York: Elsevier Science Pub. Co., 1985.
Foulis, D., C. Piron and C. Randall, 'Realism, Operationalism, and Quantum Mechanics', *Foundations of Physics* 13, no. 8. (1983): 813–41.
Frauchiger, Daniela and Renato Renner, 'Quantum theory cannot consistently describe the use of itself', *Nature Communications* 9, no. 3711 (2018): 1–10.
Friedman, Michael, 'Newton and Kant on Absolute Space: From Theology to Transcendental Philosophy', in Michel Bitbol, Pierre Kerszberg and Jean Petitot (eds), *Constituting Objectivity: Transcendental Perspectives on Modern Physics*, 35–50, Dordrecht: Springer, 2009.
Galavotti, Maria Carla, 'Operationism, Probability and Quantum Mechanics', *Foundations of Science* 1 (1995/6): 99–118.
Hawking, Stephen and Roger Penrose, *The Nature of Space and Time*, Princeton, NJ: Princeton University Press, 2010.
Hegel, Georg W. and George di Giovanni, *The Science of Logic*, Cambridge: Cambridge University Press, 2010.
Heidegger, Martin, *Being and Time*, New York: HarperPerennial/Modern Thought, 2008.
Heisenberg, Werner, *Physics and Philosophy: The Revolution in Modern Science*, New York: Harper & Brothers Publishers, 1958.
Hughes, R. I. G., *The Structure and Interpretation of Quantum Mechanics*, Cambridge, MA: Harvard University Press, 1989.
Jammer, Max, *The Philosophy of Quantum Mechanics: The Interpretations of Quantum Mechanics in Historical Perspective*, New York: Wiley, 1974.
Jeans, James, *Physics and Philosophy*, Cambridge: Cambridge University Press, 2009.
Joravsky, David, *Soviet Marxism and Natural Science, 1917–1932*, London: Routledge, 2009.
Kastner, Ruth E., Jasmina Dugić and George Jaroszkiewicz, *Quantum Structural Studies: Classical Emergence from the Quantum Level*, Hackensack, NJ: World Scientific, 2017.
Katsumori, Makoto, *Niels Bohr's Complementarity: Its Structure, History, and Intersections with Hermeneutics and Deconstruction*, Dordrecht and New York: Springer, 2011.

Lange, Marc, *An Introduction to the Philosophy of Physics: Locality, Fields, Energy and Mass*, Oxford: Blackwell, 2002.
Lefebvre, Henri. *The Production of Space*, Oxford: Blackwell Publishing, 1991.
Lenin, Vladimir I., *Materialism and Empirio-criticism: Critical Comments on a Reactionary Philosophy*, New York: International Publishers, 1970.
Levinas, Emmanuel, *Totality and Infinity: An Essay on Exteriority*, trans. Alphonso Linguis, Pittsburgh: Duquesne University Press, 1969.
Lewis, Peter J., *Quantum Ontology: A Guide to the Metaphysics of Quantum Mechanics*, New York: Oxford University Press, 2016.
Maleeh, Reza and Parisa Amani, 'Pragmatism, Bohr, and the Copenhagen Interpretation of Quantum Mechanics', *International Studies in the Philosophy of Science* 27, no. 4 (2013): 353–67, DOI: 10.1080/02698595.2013.868182.
Maudlin, Tim, *Philosophy of Physics: Space and Time*, Princeton, NJ: Princeton University Press, 2012.
Maudlin, Tim, *Philosophy of Physics: Quantum Theory*, Princeton, NJ: Princeton University Press, 2019.
Mizuno, Hiromi, *Science for the Empire: Scientific Nationalism in Modern Japan*, Stanford, CA: Stanford University Press, 2009. A broad analysis of science in Japan with a chapter devoted to its relation to Marxism as well as Tanabe.
Moore, Thomas A., *A Traveler's Guide to Space-time: An Introduction to the Special Theory of Relativity*, New York: McGraw-Hill, 1995.
Murakami, Yuko and Manabu Sumida, 'History and Philosophy of Science in Japanese Education: A Historical Overview', in *International Handbook of Research in History, Philosophy and Science Teaching*, 2217–44, Dordrecht: Springer, 2014.
Murdoch, Dugald, *Niels Bohr's Philosophy of Physics*, Cambridge: Cambridge University Press, 1987.
Ney, Alissa, 'The Status of our Ordinary Three Dimensions in a Quantum Universe', *NOUS* 46, no. 3 (2012): 525–60.
Ney, Alyssa and David Z. Albert, *The Wave Function: Essays on the Metaphysics of Quantum Mechanics*, Oxford: Oxford University Press, 2013.
Peacock, Kent A., *The Quantum Revolution: A Historical Perspective*, Westport, CT: Greenwood Press, 2008.
Perovic, Slobodan, 'Why Were Matrix Mechanics and Wave Mechanics Considered Equivalent?', *Studies in History and Philosophy of Modern Physics* 39 (2008): 444–61.
Plotnitsky, Arkady, *The Knowable and the Unknowable: Modern Science, Nonclassical Thought, and the 'Two Cultures'*, Ann Arbor: University of Michigan Press, 2002.
Plotnitsky, Arkady, *Niels Bohr and Complementarity: An Introduction*, New York: Springer, 2012.
Plotnitsky, Arkady, *Reading Bohr: Physics and Philosophy*, Dordrecht: Springer, 2006.
Polkinghorne, John C., *Quantum Theory: A Very Short Introduction*, Oxford and New York: Oxford University Press, 2002.
Rohrlich, Fritz, *From Paradox to Reality: Our New Concepts of the Physical World*, Cambridge: Cambridge University Press, 1987.
Ryckman, Thomas, *The Reign of Relativity: Philosophy in Physics, 1915–1925*, New York: Oxford University Press, 2005.

Sachs, Mendel, *Relativity in Our Time: From Physics to Human Relations*, London and Washington, DC: Taylor & Francis, 1993.
Sheehan, Helena, *Marxism and the Philosophy of Science – A Critical History: The First Hundred Years*, Atlantic Highlands, NJ: Humanities Press, 1993.
Simao, Eikoh, 'Some Aspects of Japanese Science, 1868–1945', *Annals of Science* 46 (1989): 69–91.
Stannard, Russell, *Relativity: A Very Short Introduction*, Oxford: Oxford University Press, 2008.
Stevens, S. S., 'The Operational Definition of Psychological Concepts', *Psychological Review* 42, no. 6 (November 1935): 517–27.
Walach, Harald and Nikolaus von Stillfried, 'Generalised Quantum Theory – Basic Idea and General Intuition: A Background Story and Overview', *Axiomathes* 21 (2011):185–209.
Watanabe, Masao, *The Japanese and Western Science*, Philadelphia: University of Pennsylvania Press, 1991.
Watanabe, Masao, *Science and Cultural Exchange in Modern History: Japan and the West*, Tokyo: Hakusen-sha, 1997.
Wendt, Alexander, *Quantum Mind and Social Science: Unifying Physical and Social Ontology*, Cambridge: Cambridge University Press, 2015.
Yajima, Suketoshi, 'The European Influence on Physical Sciences in Japan', *Monumenta Nipponica* 19, no. 3/4 (1964): 340–51.

Index

ability 31, 38–9, 90
absence 3, 26, 43, 117, 145
absolute 3, 4, 18, 23, 34, 42, 44–6, 49,
    51–3, 56, 59–63, 67, 71, 72,
    74–7, 79, 89–90, 93, 98, 101–2,
    114, 116–17, 120, 125, 129, 135,
    138–42, 144, 156 n.45,
    162 n.15, 163 n.24
absolutely 18, 37, 40, 61, 72–3, 79, 83,
    88–9, 96, 104, 125, 130
absoluteness 18, 117
absolutes 137
acceleration 77
actants 145
acted 26, 52
acting 15–18, 22, 26, 29, 36–9, 58–60,
    63
action 15–17, 21–2, 24–5, 31, 36, 38,
    42, 57, 75, 81, 85, 92, 120,
    143, 157 n.54, 159 n.85,
    160 n.98
active 5, 8, 15–17, 31–40, 43, 51, 56–7,
    68, 75–6, 79–81, 86, 88–9,
    92–3, 125, 135, 156 n.45
activity 11, 16, 19, 21–2, 23–31, 34,
    36–7, 117, 126
actor 75, 153
acts 19–20, 29, 31, 36–8, 40, 56–7,
    78–9, 89, 127, 136, 157 n.54
actual 20–2, 26–8, 36, 38–40, 43,
    49–50, 88, 90–1, 112, 130
actuality 39–40, 45, 97, 144
actualization 39, 99, 106
    actualize 89, 105
aesthetics 15–18, 31, 35, 38, 51–2
aether 72, 74, 79, 114, 123, 125,
    162 n.14–15, 166 n.36
    aether-filled 19
agency 5, 17, 19, 20, 24, 32–8, 51–6,
    62, 145, 102, 138, 147, 152 n.7

alienation 4, 103, 113, 126
alternating 86, 90
analytic 30, 45, 89, 91, 104–5, 122
Anthropocene 144
anthropocentric 106, 135, 142
anthropomorphism 89
anti-intellectual 128
antimatter 5, 100, 137
antimonies 114, 155 n.39
antiparticle 173
anti-Semitism 94, 164
apparatus 70, 138–9, 142, 155
apparatuses 49, 117
apperception 22
Aristotle 93, 95–6, 99, 105–8, 128,
    153 n.15
assemblage 111, 131
assemblages 48
atom 7, 81, 83, 92, 96, 101, 137
atomic 45, 63, 80–2, 96–7, 99–100,
    122, 124
    see also subatomic
atomistic 19, 75, 82, 156 n.50
axioms 116, 119
    axiomatic 85, 123
    axiomatizations 9
axis 37, 61, 71, 77, 152 n.2

Badiou, Alain 156 n.53
Barad, Karen 147
basho 場所 15–20, 35, 40, 52, 59, 61,
    92–3, 127, 136, 140–3,
    155 n.33, 156 n.49, 158 n.74,
    159 n.81, 160 n.91, 163, n.19
    basho-emplaced 57
beables 4, 138–9, 146
being-based 42
being-for-others 97
being-oriented 4
beings 51, 103, 159 n.57

being-there 68, 97
  see also Dasein
Bell, John Stewart (1928–1990) 84,
  92, 138–9, 146
binary 3, 9, 142, 145–6
biological 6, 52, 145
biologically 56
biopolitical 140
de Biran, Maine (1766–1824) 16,
  32–4, 40, 57
bodies 18, 23, 49, 51–6, 75, 137,
  160 n.93, 161 n.106
bodily 48–56, 62–3, 75, 135
body 50–4, 56, 62, 69, 75, 81, 88, 125,
  147, 167 n.45
Bohm, David Joseph (1917–1992) 84,
  146
Bohr, Niels (1885–1962) 5, 7–8, 41,
  63–4, 70, 81–6, 88, 90, 101–2,
  104, 114, 124, 137–8, 143, 146,
  158 n.72, 163 n.23
Bohr–Einstein 90
bombs 81
Boolean 100
boson 43
bourgeois 111–13, 121, 127, 132,
  168
  bourgeoisie 118
Brentano, Franz (1838–1917) 16,
  27–9, 31, 33, 37, 40, 155 n.34
Bridgman, Percy Williams (1882–
  1961) 5, 13, 41–3, 47–9, 52, 54,
  56, 58–9, 63–4, 143, 157 n.56
  and n.61, 158 n.68, n.71–2
  and n.74, 160 n.97, 161 n.103
de Broglie, Louis (1892–1987) 83–4,
  122
Buddhism 8, 19, 32, 97, 141
Buddhist 2–3, 6, 9, 11, 20, 27, 32, 40,
  58–9, 62, 69, 97, 108, 135,
  139–40, 153 n.14, 154 n.26,
  156 n.53, 157 n.55, 160 n.98

Cantor's set theory 13, 58, 60, 123,
  156 n.53

capitalism 11, 117, 121, 124, 126, 132,
  140
Cartesian 31, 52, 138
  see also Descartes, René
causal 37, 40, 47, 55, 71, 73, 119–20,
  125–6, 131, 142–3, 154 n.29,
  168
  causally-triggered 170 n.41
  causal-mechanistic 140
causality 10, 12, 22, 26, 34–5, 37, 47,
  63, 69, 73, 85, 99, 102, 113,
  125–6, 131, 143, 159 n.77,
  160 n.97, 167 n.51, 168 n.53
causation 35, 39
  causative 54
cause 19, 23, 37, 47, 71, 102, 125, 131,
  160
  causes 58, 140, 156 n.50
cells 145
censorship 111, 113
certainty 16, 31–4, 85, 117, 136
characteristics 18, 26, 56, 73, 107
chōra 15, 18–19, 143, 151 n.8,
  160 n.91
classical ([Newtonian] physics and
  mechanics) 4–6, 8, 10, 12,
  17–18, 26–7, 31, 35, 40, 42, 44,
  48, 56, 63, 67, 71–3, 75, 77–8,
  80, 82–4, 86–7, 89–90, 92–5,
  98–101, 104–7, 113, 124,
  127–8, 136, 138, 143–7,
  154 n.29, 157 n.55 and n.61,
  158 n.67 and n.72, 159 n.77,
  160 n.91 and n.97, 162 n.14,
  169 n.31, 170 n.41
classical–quantum 145
co-constitution 19
codetermination 156
co-emergence 17, 57–8, 60, 62, 135
  co-emerge 16
coexistence 4, 76, 139
cognition 6, 18, 25–6, 36, 43, 52, 57,
  75, 86, 88–9, 99, 103, 105, 131,
  155 n.34
colonial 7–9, 11, 67

colonialism 139
colonialist 12
colonies 11, 126
colonization 9, 11–12
colour 20–1, 25, 34, 38, 81, 152 n.11
combination 22, 50, 88, 99
  combinations 49, 73, 75, 122
  combining 21
combine 33, 36, 58, 91, 97, 99–100, 154
  combined 21, 31, 45, 84
commonality 31
commonplace 18, 49, 63, 156 n.54
commons 2
commonsensical 112
  see also everyday
communism 111
  communist 6, 113, 127, 168 n.54
community 18, 114, 117
commutation 75, 78–9, 83, 100, 162 n.15
complementarity 41, 63, 69–70, 84, 104, 138, 140, 144–7, 163 n.23
complete account of the physical world 13, 42, 47, 68, 70, 80, 86, 90, 92, 98
conceptualizing 72, 111, 138, 142, 156 n.45
consciousness 21–2, 25–5, 37, 39–40, 50, 52, 55, 86, 89, 94, 119–21, 142–3, 154 n.26, 155 n.33, 168 n.56
constructability 146
contents 21–2, 24–5, 28–30, 32–7, 39, 40, 44, 49, 54, 63, 72, 86, 91, 103, 123, 143
  sensory 21–2, 29, 152 n.10
contingency 2, 21, 90–1, 127, 131, 167 n.51
continuity 22–5, 27, 29, 32, 34, 47, 69, 97, 115, 123, 125, 130, 140–2
continuum 31, 77
  dialectical 155 n.34
contraction 72, 123

contradiction 27, 44, 46, 50–7, 60–3, 89, 99, 117, 123, 126, 127–9, 136, 139, 140–3, 145, 163 n.21
contraries 112, 129, 166 n.43
Copenhagen interpretation 63, 70, 138, 143, 146
co-present 168
co-productive 15, 57, 167
  co-produced 160
copy theory 56, 161
co-relation 19
corporealistic 156
  corporeally 137
correlations 144
  correlative 21, 86
correspondence 59, 82–3
cosmic 137
  time 125, 136, 166 n.36
cosmological 13, 87, 126
cosmopolitanism 9
counter-hegemonic 4
counterintuitive 16, 54
counter-narratives 117
co-variant 26, 40
creative 9, 16, 21, 32, 35, 37–8, 53, 62, 142
  creativity 6, 15, 38, 40, 140
crisis 46, 68, 71, 81, 113, 121, 127–30, 132, 167 n.43
cultural 2, 9, 111, 118, 127, 142
cultural capital 2
culture 2, 68, 112, 117, 138, 146
  cultures 6, 11, 13, 87
curvature 22, 26, 77, 93, 95

Dasein 68, 97, 113, 164 n.39
datasets 147
de-absolutized 116
debates 2, 7–9, 12, 41–2, 68–9, 90, 102, 107, 111, 124, 135, 137, 143, 147, 151 n.11, 164 n.33
Deborin, Abram Moiseevich (1881–1963) 132, 164 n.33
decidability 82, 85–6, 112

decisions 48
decision-making 145
deconstruction 13, 32, 63, 113,
    153 n.19, 157 n.55, 167 n.51
Dejima 12
democratic 113
depiction 8, 58, 83, 88, 98, 103, 129
Derrida, Jacques (1930–2004) 106
Descartes, René (1596–1650) 124,
    161 n.101
    see also Cartesian
description 27, 30, 44, 71, 80, 85, 90,
    124, 138, 158 n.72, 160 n.92,
    170 n.33
destabilization 118, 138
detection 7
determination 9–10, 20, 22, 29–31,
    44–7, 50, 52–4, 58–9, 61, 72–3,
    75–6, 78, 81–2, 85, 94–6,
    112–13, 115, 121–4, 138–45
    determinable 117
determinism 57, 73, 85, 125, 130–1,
    143, 156
dialectics 3–5, 11, 15, 20, 35, 42, 49,
    52, 57, 59, 61–2, 67, 69–71,
    78–80, 86–94, 96–7, 103–5,
    107–8, 112–13, 116–18, 120,
    122–7, 129–31, 135–7, 139–43,
    145, 147, 155 n.36, 161 n.102,
    163 n.19, 164 n.32, 165 n.15,
    166 n.35–6, 167 n.47, n.49
    and n.52, 168 n.56, 170 n.33
    see also nature dialectics;
    dialectics of nature
dialectics of nature 71, 79, 87–9, 91–2,
    124, 164 n.32
    see also nature dialectics
dimension 4, 26, 76, 77, 93, 135, 140
Dirac, Paul (1902–1984) 5, 8, 13, 28,
    62, 67, 73, 83, 93, 95–7, 99–100,
    102–3, 108, 114, 123, 137, 144
direct 17, 29, 3, 52, 70, 80, 113, 121–2,
    132, 139
discontinuity 69, 81, 95, 97, 125, 130,
    140

discontinuous 42, 81–2, 96, 103, 107
discourse 1–2, 8, 11–13, 48, 87–8, 107,
    111, 113–19, 121, 127, 135,
    156 n.53, 160 n.96, 166 n.22,
    167 n.44
discovery 5, 18, 43, 69, 80, 105, 108,
    122, 137
    discoveries 8, 11, 70, 78, 80
dogmatism 74–5, 105, 108
Dostoevsky, Fyodor Mikhailovich
    (1821–1881) 127
double-slit 69
dual 83, 88
dualisms 142
duality 67, 86, 88, 97, 139
Dutch learning (*Rangaku*) 3, 12

ecological 16, 116
ecologies 5
effects 41, 43, 58, 81, 87, 90, 119
Einstein, Albert (1879–1955) 7–8, 16,
    18, 26–7, 40, 42–5, 70–1, 73–4,
    76–8, 81, 90, 92, 94, 114, 120–2,
    143, 157–8
    see also Einstein's special theory of
    relativity
Einstein–Copenhagen rift 92
    see also EPR paper
Einsteinian 17, 31, 73, 121
Einstein's special theory of relativity
    6, 10, 26, 38, 42–3, 45, 71, 73–8,
    82, 129, 152 n.2
electromagnetic 45, 96, 98, 122
electromagnetism 18, 123
electron 45, 81–7, 96–9, 101, 138, 144,
    157 n.59
elementary particles 7, 38, 43, 67, 79,
    137
embodiment 16, 58–9, 62–3, 118, 147
emergence 11, 18, 22, 26–7, 57–9, 61,
    63, 69, 78, 108, 135, 137, 139,
    141, 144–5
emotions 38–9
empathy 31, 155
emperor 12

empires 9, 11
empirical 3, 18–19, 41–3, 74, 77, 96,
    99, 105–7, 116, 118, 138,
    157 n.63, 158 n.69, 162 n.13
empiricism 42, 49, 74, 158
Empirio-criticism 124
emplacement 4, 118, 136, 140
emplacement-orientation 144
emptiness 3–4, 27–8, 39, 97, 139–41
encyclopaedic 63, 111
energy 13, 18, 26, 30–1, 44, 47, 67, 73,
    80–2, 84–5, 95, 97–9, 102,
    122–3, 144
enlightenment 4, 116
entanglement 41, 137
environment 57–60, 70–1, 137
epistemes 5, 147
epistemology 3, 6–9, 17, 41, 43, 57, 63,
    69, 81, 86, 88, 92, 106–7, 115,
    118, 121, 135, 137, 139, 143,
    160 n.96, 162 n.13, 167 n.52
EPR paper 42, 90, 145, 158 n.72
equations 13, 44–5, 83–5, 97, 100,
    157 n.61, 162 n.14
equivalency 22–3, 26–7, 42, 55, 84, 99,
    123
essentialism 8, 105, 108, 139
ethics 2, 140, 145–6
everyday 10, 42, 53–4, 56, 87, 112–13,
    119, 121–2, 145
Euclidean 19, 26, 45, 77, 116–17,
    143, 154 n.29, 157 n.61,
    158 n.73
eugenics 8, 67
Eurocentric 2, 9
experience 15, 17, 19, 21–2, 24, 30,
    38–40, 42, 46–50, 53, 56, 58, 87,
    90, 115, 135, 147, 152 n.11,
    157 n.59, 158 n.66 and n.72,
    159 n.76
    experiential 21–2, 42, 44, 46, 63,
    119
experiment 28, 41, 46, 56, 63, 69, 72,
    100–1, 145, 158 n.66 and n.71,
    170 n.76

experimental 10, 42, 46, 48–9, 56,
    69–70, 82, 96, 99, 144, 147,
    157 n.61, 170 n.41
experimentation 43, 68, 108, 135
experiments 8, 19, 41–2, 44–7, 49, 56,
    63, 72, 76, 80, 90, 99, 103, 105,
    146–7, 154 n.27, 158 n.71,
    159 n.76
exploitation 113, 125, 126
exploration 6, 26, 43, 68, 136, 139,
    158 n.65
expression-agential 53
expressive 16, 51–5, 57, 62, 142
extrinsicality 141

fact-act (*Tathandlung*) 16, 27–8, 40,
    135, 155 n.32, 156 n.45
facts 21, 27, 29, 32, 49, 57–8, 63, 74,
    82, 99, 130–1, 144, 167 n.47
factual 28–9, 33, 34, 119
failure 92, 101
fallacy 9, 42, 89, 168 n.56
fascism 87
field of physics (as a science)
    111–12
fields 3–4, 15–16, 18–20, 30, 36, 45,
    62, 79–80, 93–8, 116, 123, 137,
    139, 142–3, 155 n.33–4,
    156 n.49, 158 n.68, 160 n.91,
    163 n.24
force 10, 12, 16–21, 24, 26–32, 34, 36,
    38, 47, 51, 73, 78, 95–6, 98, 116,
    123, 125, 137, 139, 141–3,
    155 n.39, 159 n.77
    force-relations 27, 140
formalism 104–5, 107–8, 123
form–content 39
form-matter 152
formulae 7, 45, 56–7, 59, 67, 84, 95,
    100, 107, 128, 138, 144–5,
    166 n.36, 167 n.49
    formula-orientation 69
four-dimensional 16, 26, 77–8, 93, 95,
    152 n.2, 155 n.40
    fourth-dimensional 77

frames 6–7, 9, 16, 59–60, 63, 100, 108, 136–7, 145–7, 158 n.65
framework 3, 46, 93–4, 127–8, 136, 141, 143
Frauchiger 145–6, 170
future 2, 29, 43–4, 47, 50–1, 53, 55–6, 61, 67, 75–6, 103, 124, 128

Galilean 70, 72
gravity 18, 45, 78–80, 94–6, 123

habits 29, 33, 40, 57, 167
Hamiltonian 100
Hegel, Georg Wilhelm Friedrich (1770–1831) 5, 16–17, 20, 35, 63, 78, 91–2, 97, 136–7, 139–42, 154 n.26, 160 n.100, 162 n.17, 168 n.56
hegemony 3, 7, 9, 42, 119, 170
Heidegger, Martin (1889–1976) 6, 8–9, 68, 87, 94, 113, 151 n.8, 163 n.19, 164 n.39
  Heideggerian 6, 92, 94, 97, 104, 114
Heisenberg 67, 70, 79, 82–5, 88, 100, 102, 104, 114, 123, 126, 137–8, 146–7
Heisig, James 139, 141
hierarchical 7–9, 142, 156
Hilbert, David (1862–1943) 123
  Hilbert space 145, 154 n.27, 155 n.44
historiographical 48, 108, 125, 159–60
homogeneity 4, 25, 47, 71, 95, 117, 119
humanism 12–13, 106
human–non-human 153
human-scaled 63
  -world 4
Huygens, Christiaan (1629–1695) 122–3
hylomorphism 152

idea 16–17, 20–1, 28, 30, 35, 48, 58, 61, 88, 91, 97, 104, 117, 139, 143, 153 n.12

ideas 12, 30, 57, 67, 81, 119, 122, 143, 156 n.54, 167 n.44
ideal 35, 37, 39, 40, 59–61, 117, 140
idealism 8, 59, 92, 94, 96, 118, 121–2, 131, 141, 155 n.32, 167 n.48, 168 n.56
idealist 78, 102, 114, 121, 126–7, 159 n.81
idealist-materialist 79
idealization 138
identical 39, 46–7, 72, 89, 105
ideology 13, 111–19, 121, 124, 127–8, 132, 136
illusion 32–3, 32, 137, 167 n.47
image 36–7, 39, 49–50, 74, 76, 88–9, 145, 159, 161, 169–70
imagination 40, 48, 52, 83, 120
imaginary and real numbers 77, 96, 164 n.44
immanent 121, 152, 154
immediacy 4, 116
impermanence 27
incompleteness 82
individual 21–6, 29, 31–5, 51, 53–4, 61, 91, 153 n.15, 159 n.86
  individuality 147
indivisibility 95–7, 99
infinitesimal 72, 146
infinity 24, 29, 38, 58, 98, 123, 156 n.53
instrumental 140
instruments 49, 75
intensity 30–4
intentionality 135
interaction 3–5, 16, 18–19, 24, 78, 85, 139, 142, 147, 160 n.92, 168 n.56
interdependencies 20, 118
interference 83
interior 25–9, 63, 121, 127, 132
interpenetration 20, 91, 99, 119
interrelation 19–20, 28, 143, 154 n.28, 164 n.32
intersubjectivity 19, 42, 104, 139, 146

Index   195

intertextuality 106, 108, 111, 118–19, 139
introspection 27, 30, 36
intuition 4, 5, 15, 17, 31, 34–5, 37, 39–40, 43, 48, 56, 62, 92–3, 100, 114–20, 122, 144, 146–7, 154 n.26, 155 n.39, 167 n.43
intuition-space 119, 120
invariants 26, 31, 59, 74, 90
   invariables 91
   invariance 95

judgement 23–5, 29, 38, 118, 143

Kant, Immanuel (1724–1804) 5–6, 15–17, 21, 30, 35, 94, 114–18, 120, 122, 135–6, 139, 142, 162 n.13
Kantian 5, 20, 43, 52, 78, 114, 116–20, 138, 141, 147, 152 n.10, 154 n.20 and n.26, 155 n.32 and n.39, 157 n.54, 162 n.13, 167 n.43

laboratory 7, 10, 42, 145
   laboratory-controlled 159
Lefebvre, Henri 115–16
Leibniz, Gottfried Wilhelm von (1646–1716) 53, 142, 154 n.21
Lenin, Vladimir (1870–1924) 124
Levinas, Emmanuel (1905–1995) 4, 163 n.19
Lewin, Kurt (1890–1947) 57–8, 160–1
life-space 55
locale 1, 70, 75–6, 79
localization 1
location 1, 55, 57, 74, 84, 87, 100, 138, 155 n.33
Lorentz, Hendrik (1853–1928) 74
   length contraction 72
   neutrality of theory 82
   transformations 74, 97
   [??] 163 n.24
Lotze, Rudolf (1817–81) 17, 20, 26

Mach, Ernst (1838–1916) 82, 163 n.25
macrological 95, 145, 147, 155 n.44, 170 n.41
   see also macroscopic
macro-physicality 146
macroscopic 82, 84, 94, 96–7, 99–100, 107, 145
   see also macrological
Maraldo, John 16, 151 n.11, 152 n.7
Marx, Karl Heinrich (1818–1883) 124, 126
Marxian 63, 70, 87, 116, 121, 128, 136, 166 n.36, 167 n.43
Marxism 90, 94, 111, 114, 126, 131, 136, 165 n.15
Marxist 5, 17, 68, 88, 92, 94, 107, 111–15, 117–18, 124–8, 132, 136, 165 n.15
   classical 166 n.35
mass 18, 26, 30–1, 47, 73, 79–80, 84, 96–7, 102, 143
materialism 5, 8, 17, 59, 70, 92, 96, 111–12, 121–2, 124, 126, 128, 131, 135, 137, 141, 166 n.36, 167 n.48
mathematical 28, 48, 55–6, 69, 71, 76, 89, 94, 98–100, 104–5, 130, 146, 161 n.101
mathematics 8, 55, 58, 77, 79, 105, 112, 121, 123, 130, 162 n.15
matrix 13, 15–16, 18, 62, 67, 69, 82–4, 100, 128, 143–7, 160 n.91
measurement 44–5, 47, 56, 72–5, 78–9, 92, 139, 147, 158 n.67, 162 n.14, 164 n.39
   measurable 80, 157 n.59
mechanical 37, 42, 48, 51, 67, 93, 95, 96, 99, 101, 120, 123, 125, 131, 144
mechanics 5, 6, 10, 13, 18, 28, 30, 31, 41–4, 46, 48, 62, 67–70, 72–3, 79–80, 82–6, 89–90, 95, 97–100, 102–4, 108, 114, 118–19, 122, 124, 126, 128,

136–40, 142, 144, 146–7, 155 n.44, 160 n.97
mediation 27, 31–2, 53, 59–60, 67, 75–6, 78, 92–5, 99, 103, 105–7, 114, 130, 144–5
mediator 4, 58–9, 61, 71, 74, 88, 90–1, 102
medium 72, 90–1, 141
mentalism 59
micrological 95
  see also microscopic
micro-probabilistic 144
microscopic 82, 85–7, 96–7, 99–100, 102, 107
  see also micrological
mimesis 145
mimicry 2
mind 23, 39, 63, 102, 112, 119
Minkowski, Hermann (1864–1909) 8, 16–18, 25–7, 40, 58, 60, 71, 76–8, 152 n.2, 154 n.29, 160 n.95, 163 n.19
modality 24, 30–1, 33, 59
monad 21, 54, 142, 154 n.21
movement (in physics) 20–3, 37, 39, 45–7, 72, 74–7, 82, 93, 95–7, 128–9, 143, 167
  rotational 95
  bodily 49, 116
  (in Nishida) from made to the maker 52–6, 62, 74–7
  (in Buddhism) movement and stillness 157 n.55
  toward substance versus toward nothingness 142
  (use in Tosaka) 167 n.43
mutuality 7, 9, 15, 18, 21, 23, 25–6, 28, 50, 53–4, 57–9, 63, 72–7, 79, 85–6, 91, 94, 104, 108, 123, 127, 136, 138–9, 144, 162 n.13
  mutually-negating 160

nationalism 9, 68, 116
nation-building 11

nature dialectics 61, 70, 78, 88–9, 91–2, 104, 124–7, 133
  see also dialectics of nature
negation 4, 36, 49, 53, 56, 61, 63, 71, 76–7, 86, 89, 91, 94–6, 102–4, 108, 127, 129, 161, 163
  negation-incorporation 162
negative 38, 52, 77, 97–9, 102–5, 107–8, 137, 141, 144
negativity 103, 141
  negatively 144
Newton, Sir Isaac (1642–1727) 6, 18, 20, 26, 35, 40, 42, 44, 49, 73, 78–9, 81–2, 95, 98, 102, 120–2, 126, 136, 138, 143, 154 n.29
Nietzsche, Friedrich Wilhelm (1844–1900) 16, 32, 154
Nishida Kitaro 西田幾多郎 (1870–1945) 2–20, 23, 25–7, 31–2, 35, 37–8, 40–3, 48, 52, 57–69, 76–8, 86–8, 92–3, 104, 106–8, 111–18, 115–18, 127–8, 132–7, 139–45, 153 n.15 and n.19, 157 n.55–6, 158 n.65–75, 159 n.81 and n.85, 160 n.91–8, 161 n.102, 168 n.56
non-continuous 81, 83, 86
non-dualistic 140, 144
non-essentializing 140
non-existent 3, 34
non-human 5, 143, 153 n.19
non-place 114
non-present 141
non-representational 94
non-self 32, 34, 156
non-visualizable 28, 147
non-Western 4
null-dimensional 137

object 8, 11–12, 16, 19–24, 27–8, 31–2, 34–40, 45, 49–50, 52, 54, 58–9, 63, 71, 76, 81, 85–6, 88, 90–1, 93, 102–3, 113, 127, 130, 135–6, 138, 142

-centred 44
-constructs 42
-situations 140
-world 34
objectification 4, 22, 24, 25, 29, 36, 68, 120, 142
objectivity 21–2, 24, 28–30, 36, 49, 51, 86, 88, 103, 142–3, 146, 154 n.28, 156 n.48
observable 26, 38, 70, 82, 139, 144, 146
observation 7, 38, 45, 70, 74, 85, 86, 88, 92, 99, 102, 103, 106, 108, 119, 127, 137, 145
observational 44, 89, 103
observations 20, 43, 70, 82, 90, 98–100, 138, 145, 155 n.44
ontic-epistemological 48, 160
ontico-ontological 57
ontological 3, 9, 13, 18, 41, 57, 78, 94–7, 99, 104–8, 135, 137–9, 141, 157, 162, 167, 170
ontology 6–8, 13, 41–2, 63, 67, 69–70, 92, 95–6, 99, 100, 104–5, 107, 113, 136–7, 139–41, 144–7
ontologists 96
operational 43–5, 47, 49, 52, 54–6, 63, 115, 158 n.71, 160 n.93
Operationalism 5, 41–4, 59, 63, 144, 157 n.56, 159 n.81, 161 n.103
Operationalist 13, 41, 48
operators 79, 162 n.15
opposition 3, 26, 34, 39, 46, 54–5, 60, 72, 78, 81, 86–7, 91, 94–7, 99–100, 103–6, 108, 112, 117, 120, 123, 126–7, 128–9, 131, 138–9, 141–2, 143, 146, 156 n.45
optics 45
orbits 82, 101
Orientalist 7
originality 154
other-consciousness 103
otherness 106
other-power 97

paintings 34
paradigm 51, 118–19, 167 n.44
paradigms 140, 146
paradox 105, 146–7, 163 n.23
parallelism 59
parameters 69, 147, 155 n.34, 158 n.65
    discursive 119
particle 7, 10, 18, 26, 38, 42–3, 62, 67–70, 78–9, 81–4, 88, 94, 96–9, 101, 107, 122–3, 137, 144–7, 155 n.44
    particle-based 42, 69
particle-wave 84, 100–1, 114, 137, 143, 147
Pauli's exclusion principle 98
perception 7, 16–18, 25, 27–34, 36–7, 39–40, 74, 108, 162 n.13
phantom-space 117
phases 81–2
phenomena 7, 15, 17, 19, 20, 27, 29, 30, 35–7, 39, 42–4, 46–8, 49, 56–8, 62, 67, 69–70, 72, 81–7, 92, 95, 102, 105, 108, 113–14, 118–19, 122, 128–9, 135, 138–40, 146–7, 157 n.59, 158 n.65, 162 n.14, 163 n.23
phenomenology 6–7, 20, 32, 69, 78, 107, 117
photoelectric 81
photon 73, 80, 83
photonic 81
photons 69, 81, 122, 144
photon's 99
pilot-wave 84
Plato 35, 37, 95–6, 140, 143, 153 n.12
    Platonic 143, 153
Plotinus 17, 19, 35, 37, 156 n.50
poiesis 5, 15–17, 4–51, 54–8, 61–3, 92, 135, 142, 159
politics 68, 112
    political 3, 8, 68, 111, 114, 116, 118, 120, 126–8, 132, 140, 168 n.55

polity 11, 111
positive electron 97
    see also positron
positivism 41, 82, 114, 116
positivities 120
positrons 97–9
post-Aristotelian 114
post-causal 136
post-classical 160
postcolonial 3–4, 9, 11, 42, 68, 107
post-Euclidean 113
post-Hegelian 69, 144
post-human 76, 106, 153
    post-humanism 5, 12, 144
post-Kantian 117
postmodern 118
post-scientific 13
poststructuralism 114, 137, 146, 168 n.58
praxis 17, 94, 111, 136
predication 23–2, 153 n.15
premises 1, 3, 63, 83, 124–5, 136
prespatial 116
probability 4, 18, 31, 43, 48, 62, 76, 83, 85, 90, 98–100, 102–3, 107, 128, 130, 136–7, 143–4
proto-post-humanism 107
proto-relativity 154 n.29
pseudo-colonial 4
psychobiologically 58
psychology, *see* topological

quantification 6, 31, 44, 73, 80, 94, 100, 102, 159 n.77, 167 n.49
quantitative 37, 46, 56
quantum-material 139
quasi-dialectical 16, 135
quasi-physical 58

racialization 8–9, 68
radiation 80–2, 86, 97–8, 122
radioactive 18, 81–2
realism 13, 145–7
reality 17, 20–2, 28, 32, 39, 43–4, 46, 48–9, 53, 56, 63, 86, 89–90, 95, 105, 108, 118, 120, 139, 141, 145, 161 n.103
realm 18, 22, 24, 29, 34, 36–7, 146, 156 n.48
reciprocal 21, 76, 88, 105–6, 123, 138, 147
recognition 1, 4, 18, 22, 67, 70, 74, 106, 120, 137, 145, 155 n.34, 170 n.33
recognizable 43, 106, 116, 136
reflexivity 16, 18, 35, 114
relationality 4, 10, 15–16, 19, 23, 25–6, 35, 62, 73–4, 77, 79, 114, 135–8, 153 n.19
relationship 5, 10, 13, 19, 22–3, 27, 32, 36, 39, 46, 50, 53–4, 58, 72, 75, 79, 86, 96, 105–7, 112, 116–17, 119, 125, 128, 135–6, 143, 160 n.94
relative 20, 45–7, 50, 71–5, 77, 79, 87, 89, 93, 99, 129, 145, 162 n.14
relativistic 4, 18, 31, 44, 59, 62, 67, 72, 74, 76, 80, 103, 129, 137, 144, 147, 157 n.61
relativity 1, 4–10, 13, 17, 18, 26, 28, 30–2, 35–6, 38, 40–1, 43–5, 58, 59, 62, 68–82, 86–95, 97–9, 101, 103–4, 112–14, 116–17, 121, 123, 129–30, 137, 142–3, 152 n.2, 154 n.29, 155 n.33, 160 n.97, 163 n.24, 164 n.39, 167 n.46
    see also Einstein's special theory of relativity
representation 6, 17, 120, 138
    as alternative translation for 'expression' (*hyōgen*) 152 n.7, 160 n.92 and n.95, 161 n.102
    in matrices 147
    pictorial 48
    representable 31, 147
    representational 117
    of sensory contents 21
    and space 115, 120, 165 n.15
    in Tosaka Jun 115–17, 119

revolution 6, 70–1, 74, 77, 79, 124, 127, 146
revolutionary 71, 78, 80, 85–6, 97, 112, 128, 143
rotation 77, 95

scale 7, 38, 41–3, 45, 70, 73, 76, 87, 92, 94, 100–1, 126, 137, 140, 145–7, 155 n.44, 164 n.44
schematics 5, 57, 59–63, 114, 130, 135, 137, 163 n.19
Schopenhauer 32, 154 n.23
Schopenhauerian 17
Schrödinger 67, 83–5, 100, 114, 122–3, 137, 143, 145–6, 157 n.61
Schrödinger's Cat 146
as negative example 170 n.41
scientism 10, 107
self-awareness 22, 25, 27–31, 34–5, 37, 57, 76, 86, 102–3, 144
self-affirmation 61
self-centred 117
self-colonization 11
self-consciousness 14, 34–6, 39–40, 89, 93, 103–4
self-contradictory 51, 53, 56, 67, 86, 104, 141
self-determination 55, 59, 61–2, 142
self-expression 53, 55, 160 n.95
self-formation 50–1, 54–5
self-identity 50–7, 60–3, 80, 136, 139–40, 142
selflessness 11, 135–6
self-negativity 81
self-negation 61
self-organizing 159 n.87
self-power 97
self-realization 20, 102, 108
self-reflection 27, 29–30, 155
self-unity 51
selves 25, 29, 33, 35, 51–3, 62
semiotics 55
sensation 15–16, 21–2, 25, 28, 30, 32–3, 118

sense-matter 143
sensory 18, 20–2, 25–6, 29, 33, 116, 147, 152 n.10
sensorial 26, 154 n.22
simultaneity 45, 50–1, 53, 55–6, 61, 68, 70, 73–6, 78, 82, 86, 90, 97–8, 115, 139, 165, 170 n.41
Soviet 111–13, 115, 126–7, 132, 164 n.33, 166 n.36, 168 n.56
-approved 7
-centric 126
space 3–6, 13, 17–20, 22–30, 34–6, 43–4, 47, 54–7, 59, 63, 67, 70–7, 79–81, 93–5, 98, 102, 111–24, 128–9, 136–46, 154 n.26–9, 155 n.36 and n.44, 157 n.61, 158 n.67, 159 n.77, 160 n.93 and n.95, 162 n.14–15, 163 n.24, 165 n.15, 166 n.18 and n.22, 167 n.43–6
see also representation
spaceless 4, 143
spaces 3–4, 28, 42, 94, 115–16, 122, 125, 136, 140, 144–5
space-time 4, 10, 12, 16–19, 23–4, 26–8, 31, 35, 56, 58, 60, 63, 67, 73–8, 80, 82, 94–6, 98, 112–13, 117, 129–31, 138–9, 142, 147, 152 n.2, 154 n.22, 155 n.40, 161 n.103, 166 n.36, 167–8 n.52
spatiality 4, 29, 98, 116, 136, 143, 154, 165 n.15
spatialization 117, 159 n.77
spatial 24, 26–7, 29–30, 50–1, 53, 75–7, 96, 114, 116, 124, 144, 146–7, 155 n.34, 162 n.15
spatio-temporality 24, 30, 40, 70–1, 79, 90, 113, 147, 154 n.22, 162 n.14, 163 n.24
stress-energy 31
structurality 26, 40, 137
structuration 30, 36, 157 n.54
subatomic 70, 84, 94, 100–1, 119, 126, 137, 145–6, 164 n.44

see also elementary particles
subject-bound 160
subject-centred 40
subjectivity 22, 24–5, 34–6, 49–50, 54, 69, 86, 88, 90–1, 94, 102–3, 108, 142, 159 n.82, 160 n.95, 167 n.47
subject-object 15–16, 26, 34–5, 39, 48, 62, 87, 104, 106, 114, 138–9, 160 n.96
subject-objective 91
subjects 3, 7, 10, 18–19, 25, 68, 79, 88, 103, 107, 116, 142, 146, 154, 157, 168
subject-world 86
sublation 89–91, 96, 103, 117, 123, 126–7, 129–30, 141–2, 154 n.26, 155 n.34, 160 n.100, 162 n.17
substance 28, 39, 52, 67, 72, 78, 89, 94, 121, 135, 139–40, 142
substantial 37, 61–2, 127, 131, 167 n.52
substantive 126, 141
substratum 59, 95–6, 103, 105–8, 135
substrate 11, 137
supernovae 137
superpositioning 62, 99–100, 103, 139–40, 143–5, 147, 170 n.41
superstructure 114
supra-conscious 28
supra-individual 34–5
symbolic 5, 7, 57, 128, 142
symmetry 27, 58, 71, 77, 96, 98–100, 163 n.26
synchronicity 44, 72, 165 n.15
synthesis 5, 21–2, 25–8, 30, 32–3, 35, 52, 56, 86, 91–2, 95, 99, 102–4, 112–13, 119–21, 129, 132, 154 n.25
synthetic 21, 50, 118–19, 139
systems 6, 10, 17–18, 38, 47, 56, 62, 71–4, 76, 78, 84, 93, 98, 100, 102, 106, 116–17, 119, 136, 140, 144–5, 153, 162
system-specific 58

Tanabe Hajime 田辺元 (1885–1962) 3–13, 17, 28, 43, 48, 62–3, 67–109, 111–15, 117, 128, 135–9, 143–6, 163 n.19 and 27, 164 n.32–4, 169, 170 n.33
technical 50–1, 53, 67
see also technological
technique 33, 49–51, 53
see also technology
technological 1, 9, 13, 50–2, 54–5
see also technical
technology 2, 11–12, 50–1, 118, 145, 147, 159 n.82, 160 n.93
see also technique
tensor analysis 94–5
temporal 18, 20, 24, 27, 29–30, 36, 47, 50–1, 53, 69, 76, 82, 85, 95, 98, 116, 124, 154 n.22, 155 n.34, 165 n.15
thing-in-itself 35, 78, 97, 117
three-dimensionality 76–7
threshold 118
thrownness 68, 94, 113
see also Dasein
Tokugawa 12
topological 41, 55, 57–9, 61
psychology 41, 57–9, 61, 160 n.96
topos 35, 55, 159 n.81, 160 n.91
Tosaka Jun 戸坂潤 (1900–1945) 3, 5–8, 10–12, 63, 67–8, 70, 77–8, 87–8, 107–8, 111–33, 135–7, 165 n.15, 166 n.35–6, 167 n.43–52, 168 n.56
totality 87, 113
totalities 116, 118, 166
totalizing 42
transcendence 24–5, 51, 55
transcendent 19
transcendental 6, 20, 25, 35, 78, 104–5, 108, 114, 116–17, 138, 141–2
transformations 70, 72–4, 89, 97, 125, 136
transindividual 31, 33
translational motion 95

uncertainty 24, 62, 67, 79, 83–6, 90,
    100–4, 125, 131, 140, 143–4,
    146, 167
undecidability 85–6, 144
undifferentiatedness 3, 11, 15, 108,
    127, 139, 141
universal 17–18, 21–2, 24–6, 33, 55,
    58–61, 74–5, 79, 87, 91, 94–6,
    114, 118, 127, 136–7, 153 n.15,
    159 n.76, 161 n.101
  universality 21, 152 n.11
unvisualizable 146
utilitarianism 50
utopian 114, 126

vacuum 98
variables 44, 61, 84, 102, 138, 145,
    147
vectors 27, 30, 98, 100, 107, 145,
    154 n.27
  vector-depiction 82
velocity 26, 45, 47, 71–4, 84, 97, 102,
    167 n.45
visual 21, 25, 29, 116
visualizability 28, 83, 98, 146
vitalism 17, 159 n.87
void 3, 11, 15, 26, 43, 57, 63, 69, 97–8,
    108, 113, 135, 137, 139–40, 145
  -based 42, 116, 136

  -substratum 145
  -orientation 58
  -orientated 13, 117
void-holes 5, 13, 144
volitional 51, 143
voluntary 35

waves 10, 18, 26, 38, 42–3, 47, 62,
    67–9, 79, 81–4, 88, 90, 96–9,
    101, 107, 122–3, 128, 130, 137,
    144, 146–7, 157 n.59
  wave-based 44
  wavelike 123
  wave-orientation 72
  wave-particles 4, 84, 137, 157 n.59
    and n.61
wavelength 45, 84, 143
Westernization 11
world-line 15–17, 26–7, 58, 60, 76,
    137, 152 n.2, 160 n.95
world-philosophical 139
worlds 27, 36–40, 49, 55, 76, 79, 91,
    93, 95, 136
  worlding 77
world–self 59

X-rays 45, 81

yin-and-yang 13

www.ingramcontent.com/pod-product-compliance
Ingram Content Group UK Ltd.
Pitfield, Milton Keynes, MK11 3LW, UK
UKHW021833220426
470268UK00007B/137